T0188984

Advances in Oil and Gas Exploration & Production

Series Editor

Rudy Swennen, Department of Earth and Environmental Sciences,
K.U. Leuven, Heverlee, Belgium

The book series Advances in Oil and Gas Exploration & Production publishes scientific monographs on a broad range of topics concerning geophysical and geological research on conventional and unconventional oil and gas systems, and approaching those topics from both an exploration and a production standpoint. The series is intended to form a diverse library of reference works by describing the current state of research on selected themes, such as certain techniques used in the petroleum geoscience business or regional aspects. All books in the series are written and edited by leading experts actively engaged in the respective field.

The Advances in Oil and Gas Exploration & Production series includes both single and multi-authored books, as well as edited volumes. The Series Editor, Dr. Rudy Swennen (KU Leuven, Belgium), is currently accepting proposals and a proposal form can be obtained from our representative at Springer, Dr. Alexis Vizcaino (Alexis.Vizcaino@springer.com).

More information about this series at http://www.springer.com/series/15228

Niranjan C. Nanda

Seismic Data Interpretation and Evaluation for Hydrocarbon Exploration and Production

A Practitioner's Guide

Second Edition

 Springer

Niranjan C. Nanda
Petroleum Geophysicist
Cuttack, Odisha, India

ISSN 2509-372X   ISSN 2509-3738   (electronic)
Advances in Oil and Gas Exploration & Production
ISBN 978-3-030-75303-0   ISBN 978-3-030-75301-6   (eBook)
https://doi.org/10.1007/978-3-030-75301-6

Preface

The first edition of the book was published in March 2016. It was intended for postgraduate students of geoscience and young professionals engaged in the petroleum industry as an aid in understanding the fundamentals of seismic data interpretation and evaluation and their application in petroleum exploration and exploitation. For an effectual evaluation of data, a seismic analyst needs familiarity with the basics and working knowledge of various disciplines such as geophysics, geology, geochemistry and reservoir engineering. These days students and young professionals come well-armed with the basics in their respective disciplines. What this book intends to present is an organized and cogent template for a systematic and synergetic approach to synthesize the multidisciplinary data which can help in the quest for finding more hydrocarbon reserves. While the preliminaries are mostly kept to the bare minimum, the emphasis in the book has been on the interpretation workflows and practices, traditionally followed in the industry. These are briefly discussed in simple and practical ways, interspersed with ample illustrations and case study examples, along with the problems that are commonly encountered. However, my experience over the past several decades in terms of interactions with students and practicing young interpreters at work as well as the feedback received from many readers prompted me to expand some of the themes a little more explicitly and add some material anew to come out with publication of the second edition of the book.

The book was earlier structured principally with two modules, namely exploration seismic and reservoir and production seismic, primarily involving conventional reservoirs. However, with the global attention leaning more towards exploring and exploiting unconventional reservoirs, a major addition in this edition is the third module which comprises the unconventional reservoirs under the heading 'Unconventional Reservoirs Exploration and Production—The Role of Seismic'. Three additional chapters under this module focus on oil sands, heavy oil, tight oil and gas sands, basin-centred gas accumulations (BCGAs), coal bed methane (CBM), shale oil and gas, gas hydrates and fractured-basement reservoirs. These topics are discussed succinctly to create awareness. Since most unconventional reservoirs are incapable of producing hydrocarbons under primary recovery, exploitation through secondary and enhanced recovery methods is also outlined. More importantly, the role of seismic in exploring and exploiting these unconventional reservoirs and in monitoring the recovery processes for efficiency is also included.

Almost all chapters in the book are expanded with more explicit explanations and addition of new subject matter. Some of the noteworthy inclusions are the concept of rock microstructure, its physical and mechanical properties, anisotropy and building mechanical earth models (MEM). Other interesting additions are polarity display conventions linked to energy source wavelet, instantaneous velocity and velocity modelling, geochemical analysis and evaluation of hydrocarbon source and generation, tomography and microseismic surveys, depth conversion vis-a-vis prestack depth migration (PSDM), pore-pressure prediction, AVO modelling, phase rotations for zero-phase 3D data and thin-bed reflectivity inversion. The author strongly believes that graphics, sketches and image illustrations with explicit captions expound the concepts better than the descriptive scripts in texts. Accordingly, many more illustrations and case examples are included in this new edition.

Lastly, the limitations of seismic technologies and techniques are underscored with more examples of failures and pitfalls, an aspect which usually remains unreported or underreported in the industry. Pitfalls are an intrinsic part of seismic interpretation along with serendipity, though the latter is always welcomed due to encouraging breakthroughs, the pitfalls are spurned for causing setbacks. Nonetheless, it is hoped that the new edition of this book provides enough fodder to the inquisitive and sharp minds of professionals for improving their imaginative skills and work practices, which hopefully will help them excel in their quest for finding more hydrocarbon through cognitive seismic data evaluation.

Cuttack, India   Niranjan C. Nanda
March 2021

Acknowledgements

A few years ago, persistent persuasion of my good friend and erstwhile colleague at ONGC, Satinder Chopra, Arcis Seismic Solutions, TGS, Calgary, Canada, compelled me to author the book titled *Interpretation and evaluation of seismic data for Hydrocarbon Exploration and Production—A Practitioner's Guide*, which was published by Springer in March 2016. Over these ensuing years, my work association with young seismic interpreters and feedback from several readers encouraged me to cover fresh ground which may seem elementary but I believe is essential for seismic analysts to become more effectual. Once again, Satinder obliged by playing a major role in expounding several concepts through personal correspondences and most importantly in providing excellent graphics and images for illustrations used in this edition. Understandably, without his help, the second edition of the book would not have been possible. I express my indebtedness and profound gratitude to Satinder for his help and support.

I thankfully acknowledge ONGC, India, Arcis Seismic Solutions, TGS, Canada, Society of Exploration Geophysicists (SEG), European Association of Geoscientists and Engineers (EAGE), American Association of Petroleum Geologists (AAPG), Canadian Society of Exploration Geophysicists (CSEG), Alberta Geological Survey (AGS) and GeoExpro and USGS for permitting the use of several graphics and images in the book. My thanks are also due to the individual authors, Tako Koning, Ray Bosewell, Ross Crain, M. Kurihara and Jane Whaley who accorded permission for use of their illustrations in my book. I would also like to express thanks to my young friend Ravi Kumar of Saturn Energy Solutions, Hyderabad, India, who helped me with reworking and improving some of the figures and images for inclusion in the book.

I sincerely thank Springer publishers for their dexterous handling of the manuscript for print.

Finally, I wish to gratefully acknowledge the support and care my wife extended during the difficult time of the COVID pandemic. Her patience and steadfast encouragement to continue was an inspiration, for which I remain indebted forever.

Cuttack, Odisha, India   Niranjan C. Nanda
March 2021

Contents

Exploration Seismic

Seismic Wave and Rock-Fluid Properties 3
Seismic Wave and Propagation 3
Energy Losses 4
    Absorption 4
    Scattering 4
    Transmission 5
    Spherical (Geometrical) Divergence 5
Geological Significance of Energy Attenuation 5
Seismic Properties 7
Rock-Fluid Properties 7
Seismic Rock Physics and Petrophysics 7
Rock Microstructure 8
    Physical Properties of Rocks 8
Fluid Properties of Rocks 13
    Pore Fluid and Saturation 13
    Permeability 14
    Viscosity 14
Pressure 14
    Normal Pressures (Hydrostatic Pressure) 15
    Abnormal Pressures 15
    Overpressures 15
    Underpressures or Subnormal Pressures 16
Temperature 18
Seismic Rock Physics Modelling (RPM) 18
Mechanical Properties of Rocks, Geomechanics 19
    Young's Modulus 19
    Poisson's Ratio 20
    In Situ Stress 21
Mechanical Earth Modelling (MEM) 21
Limitations of Modelling 22
References 22

Seismic Reflection Principles—Basics 25
Signal-to-Noise Ratio (S/N) 26
Seismic Resolution 27

Vertical Resolution   27
Lateral (Spatial) Resolution and Fresnel Zone, Migration   30
Interference of Closely Spaced Reflections; Types of Reflectors.   34
Discrete Reflectors.   34
Transitional Reflectors.   34
Complex Reflectors   35
Innate Attributes of a Reflection Signal.   35
Amplitude and Strength.   36
Phase.   36
Frequency (Bandwidth)   36
Polarity   37
Polarity Display Conventions, Acquisition Source Wavelet.   37
Arrival Time (Onset of Reflection)   39
Velocity.   40
Seismic Display   42
Display Modes   42
Plotting Scales (Vertical and Horizontal).   44
References.   45

Seismic Interpretation Methods   47
Category I: Structural Interpretation (2D)   49
Identification of Horizons and Correlation.   49
Category II: Stratigraphic Interpretation (2D)   55
Seismic Calibration   55
Depth Conversion   59
Velocity Modelling   59
Seismic Structure Maps.   63
Seismogeological Section   65
Category III: Seismic Stratigraphy Interpretation.   65
Category IV: Seismic Sequence Stratigraphy Interpretation.   77
Low Stand Systems Tract (LST)   78
Transgressive Systems Tract (TST).   78
High Stand Systems Tract (HST)   78
Application of Seismic Sequence Stratigraphy   80
Seismic Stratigraphy and Stratigraphic Interpretation   81
References.   81

Tectonics and Seismic Interpretation.   83
Structures Associated with Extensional Stress Regimes   84
Compaction/Drape Folds.   84
Horsts and Grabens.   84
Normal Faults, Types and Compaction Faults.   84
Listric and Growth Faults and Roll-Over Structures   86
Cylindrical Faults and Associated Toe Thrusts, Imbricates   87
Structures Associated with Compressional Stress Regimes   87
Folds.   87
Reverse Faults, Thrusts & Overthrusts   88

Structures Associated with Shear Stress (Wrench) Regimes 90
   En Echelon Conical Folds. 91
   Half Grabens with High-Angle Faults 92
   Flower Structures and Inversions. 93
   Strike-Slip Faults. 93
Salt/Shale Structures 94
   Salt Diapirs—Types 96
   Significance of Seismo-tectonics in Seismic Data Evaluation 98
References. 99

Seismic Stratigraphy and Seismo-Tectonics in Petroleum
Exploration 101
Basin Evaluation 102
   Source and Generation Potential 102
   Reservoir Facies 103
   Migration, Pathways and Timing. 103
   Entrapment and Preservation. 104
Basin and Petroleum System Modelling (BPSM) 105
   Petroleum System Modelling (PSM) 106
   Geochemical Evaluation of Source and Generation Potential 106
Limitations of Petroleum System Modelling 108
Fault Attributes Analysis and Fault Seal Integrity 108
   Faults as Conduits. 109
   Faults as Seals. 109
   Fault Seal Integrity 110
Prospect Generation and Evaluation 112
Techno- Economic Analysis 112
Technical (Geologic) Risk Assessment 113
   Type of Source 113
   Migration Timing and Pathways 113
   Type of Reservoir (Clastic/carbonate) 113
   Type of Trap. 113
Estimate of Hydrocarbon In-Place (Volumetric) 114
References. 115

Direct Hydrocarbon Indicators (DHI) 117
DHI Amplitude Anomalies 119
   'Bright Spot' 119
   'Dim Spot' 119
   'Flat Spot' 120
Supporting Evidence for DHI Anomalies 122
Validation of DHI Anomalies. 125
   Angle Stack Amplitudes, Near- and Far- Stacks for Validation. . . . 125
   Delineation and Characterization of Hydrocarbon Sands 125
Limitations of DHI 128
References. 129

Borehole Seismic Techniques 131
Check-Shot Survey 132

Acquisition Layouts and Energy Source   132
The T-D Curve (Velocity Function)   133
Benefits of Check-Shot Surveys   134
Vertical Seismic Profiling (VSP)   134
Zero Offset VSP (ZVSP)   135
Source Wavelet Recording and Benefits for Seismic Data
Processing   135
Corridor Stack   136
Prediction ahead of Drill Bit   137
Offset (Non-zero) VSP   138
Walkaway VSP   139
Benefits of VSP   139
Check-Shot and VSP Survey Comparison   140
Well (Seismic) Velocity and Sonic Velocity Dissimilarities   140
Sonic Drift Correction   141
Limitations of Check-Shot and VSP Surveys   141
Cross-Well (Borehole) Survey   143
Traveltime Tomography   144
Reflection Tomography   144
Tomography Applications   145
Microseismic Surveys   145
Monitoring 'Fracking'   146
Limitations   146
References   146

Reservoir and Production Seismic

Evaluation of High-Resolution 3D and 4D Seismic Data   149
3D High Resolution Seismic   150
Data Acquisition   150
Prestack Time and Depth Migration (PSTM/PSDM)   152
3D Seismic Interpretation and Evaluation Techniques   153
Horizontal-View Seismic   154
3D Volume Interpretation and Visualization   160
3D Seismic Applications   160
Pore Pressure Prediction   161
Offshore Shallow Drilling Hazards Prediction   161
Reservoir Delineation and Characterisation   162
Reservoir Delineation   164
Reservoir Characterisation   166
4D (Production Seismic)   169
Seismic Reservoir Monitoring (SRM)   170
Reservoir Geomechanics   174
3D and 4D Seismic Roles in Exploration and Production   175
4D Seismic: Limitations   175
References   176

Shear Wave Seismic, AVO and Vp/Vs Analysis 177
Shear Wave Properties - Basics. 178
   Polarization of Waves 178
Shear Wave Data Acquisition, Processing and Interpretation 179
Mode Converted Shear Waves, P-SV, P-SH and P-SV-P Waves 180
Multicomponent Surveys (3C and 4C) 182
Benefits of Shear Wave Studies 183
   Imaging Below Offshore Gas Chimneys 183
   Validating DHI (P-wave Amplitude Anomalies) 183
AVO Analysis for Validation of Hydrocarbon Sands 185
   AVO Amplitude Analysis. 185
   Gradient and Intercept Analysis. 189
   Poisson's Ratio, Angular Reflectivity in AVO Analysis 191
   Case Example 195
   Anisotropy Effect in AVO Analysis 195
   AVO Modelling, Elastic and Zoeppritz's Gather. 197
   Limitations of AVO Analysis 198
Vp/Vs Analysis, Prediction of Rock-Fluid Properties. 199
   Vp/Vs for Fluid Prediction 200
   Case Example 200
   Vp/Vs for Lithology Prediction. 201
   Vp/Vs for Porosity and Clay Content Prediction. 202
Shear Wave Birefringence, Fracture Prediction. 202
   Limitations of Fracture Prediction 203
References. 203

Analysing Seismic Attributes 205
Primary Attributes. 206
   Tuning Thickness: Thin Bed Definition and Amplitude
   Variation 206
   Complex Trace Analysis—Amplitude, Frequency,
   Phase Polarity Sweetness. 208
   Spectral Decomposition (AVF) 211
Geometric Attributes. 212
   Dip and Azimuth. 212
   Curvature. 213
   Coherence 214
   Ant-Tracking. 217
Multiattributes Analysis and Composite Coloured (Overlay)
Displays 218
   Multiattributes Analysis. 218
   Composite and Overlay Colour Displays. 218
   Multiattributes, Applications and Significances 218
Limitations in Attribute Studies. 221
References. 222

Seismic Modelling and Inversion 223
Physical and Numerical Modelling 224
Seismic Forward Modelling 225
    1D Modelling, Synthetic Seismogram 225
    2D/3D Modelling, Structural and Stratigraphic 227
Limitations of Forward Modelling 229
Inverse Modelling Types 230
    Operator-Based Inversion 231
    Recursive Inversion 231
    Model-based Inversion 231
    Geostatistical (Stochastic) Inversion 232
Seismic Inversion, Types of Attributes 232
    Interval Velocity Inversion (Vp) 232
    Acoustic Impedance Inversion (AI) 233
    Elastic Impedance Inversion (EI) 234
    Simultaneous Inversion (SI) 235
    Density Inversion 236
    AVO Inversion 236
    Thin-bed Reflectivity Inversion 237
Phase Rotation to Harmonize Seismic Data 237
Seismic Inversion in Reservoir Modelling and Management 239
Limitations of Inversion 240
References 241

Seismic Pitfalls 243
Seismic Technology (API) Related Pitfalls 244
    Acquisition Related Artefacts 244
    Processing Linked Snags 245
    Interpretation Linked Pitfalls 246
Natural System-Related Pitfalls 252
    Wave Propagation Complications 252
    Anisotropy 254
    Time Domain Data Recording 254
    Geological Impediments, Vagaries of Nature 254
    Case Examples, Vagaries of Nature 255
Seismic Limitations 256
References 257

**Unconventional Reservoirs Exploration
and Production-Seismic Role**

Oil Sands, Heavy Oil, Tight Oil/Gas, BCGAs and CBM 261
Oil Sands, Heavy/Extra Heavy Oil 262
    Oil Sands 262
    Heavy/Extra Heavy Oil 263
    Tight Oil/Gas 265
Seismic Role 265
    Cross-Well Tomography 266

Basin-Centered Gas Accumulations (BCGAs). 266
   Types of BCGAs. 266
   Abnormal Pressure and BCGAs Reservoirs. 267
   Role of Seismic. 268
Coal-Bed Methane 269
   The Coal Reservoir. 270
   Methane Production. 272
   Seismic Role. 273
   Limitations 275
References. 276

Shale Oil and Gas, Oil Shale and Gas Hydrates. 277
Shale Oil and Gas. 278
Shale Reservoirs 278
   Shale Prospect Appraisal. 278
   Shale Oil/Gas Production 280
Seismic Role. 280
   Shale Characterization. 281
   Microseismic Monitoring of 'Fracking'. 283
Oil Shale. 283
Gas Hydrates. 284
   Oceanic Gas Hydrates. 284
   Occurrences. 285
Gas Hydrate Petroleum System. 286
   Source. 286
   Reservoir. 286
   Trap. 287
   Migration. 288
Methane Production from Hydrates. 288
Seismic Role. 289
References. 291

Fractured-Basement Reservoirs. 293
Faults and Fractures, Tectonics and Basement Rocks 294
   Faults and Fractures 294
   Tectonics. 294
   Basement Rock Types. 295
Hydrocarbon Accumulation in Fractured-Basement Reservoirs. 295
   Source. 295
   Reservoirs 296
   Traps. 298
   Reservoir Characterization-Fracture Geometry. 298
   Reserve Estimate and Productivity 299
Role of Seismic. 299
   Seismic Mapping of Faults and Fractures 299
Seismic Limitations in Mapping Basement Fractures 301
Exploring Basement Reservoirs. 302
References. 304

Exploration Seismic

Seismic Wave and Rock-Fluid Properties

Abstract

Seismic rock physics is study of seismic response of rock and fluid properties, used essentially to help predict reliably the lateral and vertical variation in rock properties in the subsurface from seismic data. This would entail knowing basics about seismic wave and its propagation in the earth and the different rock and fluid properties that impact seismic properties. Seismic waves are elastic waves and suffer loss of energy of different kinds while propagating through rock layers having different rock properties in the earth. The propagation mechanisms, the losses and their geologic significance are described.

The intrinsic seismic properties are the amplitude and velocity which are influenced by properties of rocks through which the wave travels. The elasticity and density of rocks primarily determine the seismic amplitude and velocity, though several other rock and fluid properties such as porosity, texture, fractures, fluid saturation, viscosity and factors such as pressure and temperature also affect the seismic properties. The microstructure of rock, its elastic, physical and geomechanical properties and their seismic responses, known as seismic rock physics studies are deliberated.

Descriptions of homogeneous, heterogeneous, isotropic and anisotropic rocks and seismic anisotropy, Gassmann's equation and fluid saturation, normal, abnormal and pore pressures are dealt to elaborate the seismic rock physics studies. Seismic rock physics modelling (RPM) and mechanical earth modelling (MEM), their utilities in seismic rock physics application along with limitations are mentioned.

Focusing on geologic interpretation of seismic data before introducing fundamentals of seismic principles and rock physics can be something like putting the cart before the horse. Therefore, this chapter is a revisit to the basics of seismic wave propagation and related rock physics. It answers briefly some of the important questions, as given below, which ultimately guide interpretation.

- *How do seismic waves propagate through rocks?*
- *How is seismic energy attenuated?*
- *What are fundamental wave properties?*
- *What are rock-fluid properties and how do they affect seismic response?*

Seismic Wave and Propagation

A seismic wave is an elastic wave traveling through a solid rock. When a rock is subjected to a pressure wave, its particles get displaced, transferring energy to the adjacent ones causing a seismic wave to propagate onwards in the rock through particle motions. There are two types of seismic body waves that travel in solid rocks; longitudinal (primary or compressional) waves and transverse (secondary or shear) waves. In fluids, however, only the longitudinal waves can travel.

A seismic wave propagating in the earth encounters several discontinuities (boundaries) between rock types of different physical properties and produces phenomena such as reflections, diffractions, absorptions, scatterings and transmissions (refractions). At each boundary or interface between two different types of rocks, a part of the incident energy is reflected back to the surface and the rest of energy is transmitted to the underlying rocks. Seismic methods for exploration of hydrocarbons mostly use the reflected energies of primary or compressional waves returning to the surface. Shear waves reflections are also recorded but are used in specific cases, to provide valuable support to subsurface information. Chapter "Shear Wave Seismic, AVO and Vp/Vs Analysis" (Shear Wave Seismic) provides more detailed discussion on shear seismic. As the wave energy (seismic pulse) travels downwards in solid media, it undergoes gradual loss of energy (attenuation) depending on the rock-fluid properties. Attenuation, a natural phenomenon, comprises of several types of losses and understanding the process behind each loss can be useful in interpreting the rock type.

Energy Losses

Absorption

The seismic source wave, generated at the surface, as stated earlier, propagates through a rock by transferring energy from one particle to another. In the process, a part of the energy is attenuated due to conversion of mechanical energy to heat energy through frictions at grain contacts, cracks and fractures and fluids present in pores of a rock. The frictional loss, primarily due to motion between rock particles at the point of grain contacts, is known as absorption.

Frictional loss is also sensitive, to a lesser extent, to fluid properties like saturation, permeability and viscosity as the wave travels through sedimentary rocks which are generally saturated with fluids. Absorption in rocks is related to the first power of frequency whereas in liquids it is related to square of the frequency (Anstey 1977)

Absorption is called anelastic attenuation which is frequency selective and cuts out higher frequencies progressively from the source pulse as it travels down. This results in reduced energy with a wavelet of lower frequency and lower amplitude at deeper depths (Fig. 1). Absorption effects are severe within shallow weathering zones and decrease with depth. Magnitude of absorption (friction) loss in a hard rock is liable to be much higher than that in a fluid saturated rock as friction in fluid, which is a slushy medium, is likely to be marginal (Gregory 1977). For instance, seismic data in offshore deep waters hardly ever show low- frequency domination which support that little or no energy is lost due to absorption in water column. However, there can be some absorption loss in partially saturated hydrocarbon reservoirs due to viscous motion between the rock and the fluid during the wave propagation.

Scattering

Scattering loss is a frequency dependent *elastic* attenuation linked to dispersion, a phenomenon in which velocities in rock measure differently with varying frequencies. Scattering losses are irregular dispersions of energy due to heterogeneity in rock sections, and are usually considered as apparent noise in seismic records. Scattering and absorption losses are sometimes referred to as attenuation. Geological objects of very small dimensions tend to scatter wave energy and produce diffractions rather than continuous reflections. Highly tectonized shear zones with faults and fractures, very narrow channels, pinnacle mounds etc., are some of the geologic features, most prone to scattering effect.

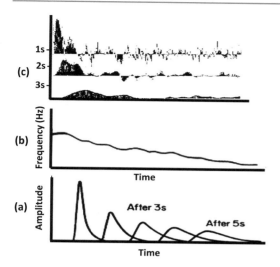

Fig. 1 Schematic showing energy loss due to absorption during propagation of a wave (**a**) time domain showing lowering of amplitude and frequency (wave-width broadening) with time (**b**) frequency domain showing loss of high frequencies progressively with time and (**c**) the over all look of a seismic trace with time (modified after Anstey 1977)

Transmission

Transmission loss is loss of energy the wave undergoes at every lithologic boundary, as a part of the energy is reflected back to the surface allowing less to go deeper. The energy loss at depth thus depends on the type and number of reflecting interfaces. It is sometimes believed that a strong reflector like a limestone or an intrusive body reflects most of energy upwards and transmits less in the process, causing poor reflections or shadows below. However, Anstey (1977) has demonstrated that strong reflectors may not be the sole reason for large transmission losses. Instead, such effects may be caused due to large number of thin layer interfaces, which even with small reflectivity but alternating signage of contrasts can create as many reflections to account for energy loss.

Transmission losses reduce amplitudes at all frequencies and are not frequency selective as in absorption. One positive spinoff of wave transmission through several thin beds can be the causal peg-leg multiples from several thin beds

which through constructive interference can create considerable reflection amplitudes to be noticed on seismic. Peg-leg multiples are intrabed short-path asymmetrical multiples generated from thin beds within a formation (Fig. 2). However, addition of several reflections tends to lower the frequencies, giving an appearance of a pulse similar to the absorption effect. Primafacie, it may be, hard to distinguish the effects on a seismic pulse due to absorption and transmission losses.

Spherical (Geometrical) Divergence

Seismic wave, ideally considered travelling in the form of spherical wave front, suffers from reduction of energy as it continually moves away from source and spreads through the subsurface rocks with time (distance). This is also known as geometrical loss as it is linked to the wave-path geometry. The decay is dependent on distance from the source and increases with higher velocities due to greater distance travelled (Fig. 3).

Geological Significance of Energy Attenuation

Large attenuation losses in rocks, besides the amplitude, also lower the frequencies of seismic wave which lead to show lower velocities due to dispersion effect. Measurement of both attenuation and velocity can therefore provide complimentary information about the rock and fluid properties. Further, attenuation affecting the frequency and the amplitude content of the wavelet also results in changing the seismic wave shape. Analysis of propagation loss in rocks from the resulting changes in wave shapes can then lead to important geological information about rock and fluid properties. Some significant geologic conclusions from analysis of attenuation effect can be as below.

- Indication of high energy loss considered owing to absorption, may give a clue to the

Fig. 2 Schematic illustrating peg-leg multiples, (**a**) multiples from within and (**b**) from bottom of a thin layer, embedded in a formation. Note the typical asymmetrical short-path multiple in (**a**) which impacts greatly the reflection quality when multiple thin beds are involved

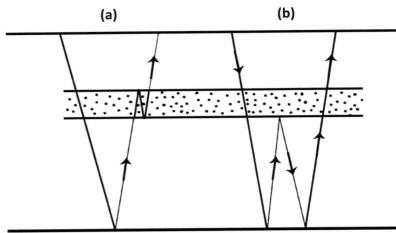

(a)   **(b)**

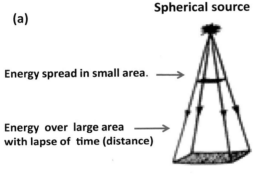

(a)

Spherical source

Energy spread in small area. $\longrightarrow$

Energy over large area $\longrightarrow$
with lapse of time (distance)

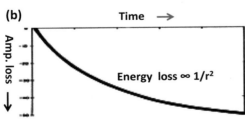

(b)   **Time** $\longrightarrow$

Amp. loss

Energy loss ∞ $1/r^2$

Fig. 3 Loss due to spherical divergence during seismic wave propagation. (**a**) spreading of spherical wave causes loss as the energy is distributed over larger area with passage of time and (**b**) is proportional to distance travelled (modified after Anstey 1977)

- Seismic evidence of high transmission loss can be suggestive of a formation consisting of cyclically alternating impedance contrasts such as in multiple thin sand layers occurring with intervening shale layers, the *cyclothems*, often typically deposited in deltaic environment. *Cyclothems* are potentially important geological plays that are commonly sought after by the explorationists.

- Scattering losses due to heterogeneity in strata may provide clues to order of irregularities in reservoirs suggesting rapid facies change such as in continental depositional environments. Similarly, scattering losses resulting in poor to no seismic reflections may indicate presence of fault and fracture zones, *mélanges* in highly tectonized zones of subduction. Seismic survey in such areas would need suitable planning of acquisition and processing techniques to achieve better seismic images.

However, types of energy losses are difficult to distinguish and determine in real field situations. Can the losses due to absorption be distinguished from those due to transmission, which cause similar effect on a wave pulse? This can be answered to some extent during data processing workflow, but usually the interpreter has little time or access to dig into data processing, that also requires special efforts to identify and quantify losses. Nevertheless, under certain favorable situations, such as in known geologic

type and texture of the reservoir rock. Unconsolidated, fractured, and poorly-sorted rocks having angular grain contacts are likely to have considerable friction (Anstey 1977). On the other hand, rocks, well-sorted and with well-cemented pore spaces will show less loss due to absorption.

areas, relatively shallow targets of exploration, in high resolution offshore marine data, it may be possible to detect some of the losses through special processing techniques. This assists in interpreting type and texture of rocks, albeit, qualitatively.

Seismic Properties

Seismic response to rock-fluid properties consists of the important intrinsic properties of seismic wave and is indispensable in the framework of exploration seismic technology. The primary properties of a seismic wave are (1) the seismic amplitude of the wave and (2) the velocity of the wave. Seismic amplitudes are the particle velocities measured by the geophones on land or the acoustic pressure by hydrophones in marine streamer surveys and velocity is with which the wave passes through the rocks. Particle velocity conveys the magnitude of the seismic disturbance (micrometers/sec) whereas wave velocity conveys the speed of the seismic disturbance at which it travels (km/s). Amplitude and velocity are the two seminal seismic properties that constitute the response and differ over a wide range, dependent on rock-fluid properties.

Seismic wave propagation in subsurface and its attendant effects brings out vital geological information about different types of subsurface rocks and their fluid contents. The rock-fluid properties, which are many, affect seismic response and can be intricately complex to decipher. Fortuitously, most of the rock-fluid properties influence one way or another the two primary physical properties of a rock, the elasticity and the density, which determine the seismic responses by amplitudes and velocities of waves. The amplitude of a seismic wave, is a function of contrasts between two impedances (a product of velocity V and density ρ) of rocks at an interface. Seismic velocity (V), on the other hand is a function of elastic modulii (E_m) and density (ρ), expressed by the equation

$$V = \sqrt{(E_m/\rho)},$$

Both compressional (P) and shear (S) velocities are influenced by rock properties albeit differently. However, the rock-fluid properties affecting mostly the P-seismic properties are discussed here while the S-seismic properties are dealt later in Chapter "Shear Wave Seismic, AVO and Vp/Vs Analysis" (shear seismic).

Having introduced the seismic properties, what are the rock-fluid properties that directly or indirectly affect the seismic properties?

Rock-Fluid Properties

The rock properties may be considered of two kinds, the physical rock properties and the mechanical rock properties, the latter mostly used for engineering purposes. Physical properties of rocks in the context of hydrocarbon exploration, also include properties of fluids that occupy rock pores. These are essentially the elastic modulii, density, porosity, anisotropy, fluid type and saturation and factors such as pressure and temperature, which impact seismic response. Mechanical properties, on the other hand, are the strength, stiffness and toughness of rocks in response to applied stress which can deform or induce changes in behavior of rock and cause changes in seismic response are studied under Geomechanics, described later in the chapter.

Seismic Rock Physics and Petrophysics

Seismic rock physics is the link between the rock and seismic properties, the cause and the effect. Knowledge of seismic rock physics provides a unique advantage in that seismic data can be used as predictive models for estimate of rock properties in petroleum exploration and production applications. Seismic rock physics may not be mixed up with petrophysics as there are several points of differences. Seismic rock physics is primarily utilized by geoscientists while petrophysics is used by log analysts. While seismic interpreters mostly use sonic and density logs to estimate rock properties, the petrophysicists use all kinds of log data in the complete log suite, together with core and production data. Whereas the seismic interpreter typically requires the logs for rock property evaluation, the log analysts may not need seismic data for the purpose. More prominently, petrophysics is a more elaborate and finer study of rock properties done in microscopic scale (mm-cm) in contrast to seismic studies in macroscopic scale (10–100 m). Consequently, seismic rock physics modeling (SRM) from seismic analyst point of view can be quite different from Petro physicist's angle.

Rock Microstructure

Since rock and fluid properties determine the seismic response, inversely, they can be interpreted from seismic data. In this context it is important to elaborate microstructure of a rock whose elements individually impact seismic properties. A rock is essentially described in terms of its framework and matrix, cement, pore space and fluid content. The framework or the skeleton comprises the coarse grains of detritus sediments whereas matrix is the very fine filling material in the space between the framework grains. Cement, on the other hand, is a secondary mineral that forms after deposition and during burial of rock that binds the framework grains together. The other important element in the pore space, the void available within a sedimentary rock, saturated with fluid (Fig. 4). All the elements that constitute a rock ultimately influence elasticity and density of a rock, and it is expedient to consider the effect of each individual element of the rock and fluid properties separately on elasticity and density to conclude their net impact on seismic response. Though the elastic moduli of rocks depend predominantly on the moduli of rock matrix, pore geometry and the elastic moduli of pore fluids, the impact of other

elements, comprising the rock microstructure are small can yet impact response considerably if they are combined to add.

Many other factors besides the rock and fluid properties such as pressure, temperature and anisotropy also considerably affect seismic properties. From the large number of rock and fluid properties that influence seismic properties, we restrict our studies limited to the usually the important ones as below that cause perceptible seismic response.

Physical Properties of Rocks

Elasticity and Elastic Constants

Elasticity of a rock is defined as the resistance it offers to stress. There are three principal elastic modulii, namely Young's modulus (E), bulk modulus (k) and shear modulus (μ). In homogeneous isotropic media, simple relations exist between these three and can be determined if any two modulii are known. Other elastic modulii such as Poisson's ratio (σ) and lambda can be determined from the three principal modulii. Lambda and Mu are known as Lame's constant. Lambda is considered a fluid indicator and can be derived from bulk (K) and shear modulii (μ) but its application is less common. The Young's modulus (E) and the Poisson's ratio (σ) and their applications are discussed under mechanical properties later in the chapter.

However, the two principal elastic moduli controlling seismic responses are the bulk modulus (k) and the shear modulus (μ). Bulk modulus is a measure of a rock's resistance to change in volume, its incompressibility and shear modulus, also known as modulus of rigidity, is the measure of resistance to deformation by shear stress, its rigidity. Depending on the type of wave, compressional or transverse, specific elastic modulii play the dominant role in determining the seismic velocity. In case of compressional waves (P-waves), both bulk modulus and shear modulus control seismic velocity and for shear or transverse waves, only the shear modulus plays the dominant role in controlling S-velocity. In an isotropic media, compressional

Fig. 4 A schematic showing the micro structural elements of a sedimentary rock: the framework comprising coarse grains, the matrix (very fine interstitial fills), cement formed post deposition (diagenesis) that binds the grains together and the pore voids filled with fluid. Each of these elements contribute to the total impact the rock creates on seismic response

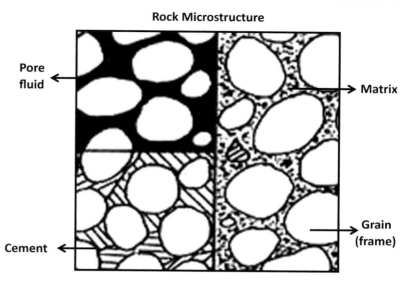

and shear wave velocities are given by the equations

$$Vp = \sqrt{[(k + 4/3\mu)/\rho]}, and$$
$$Vs = \sqrt{[\mu/\rho]},$$

One simple way to comprehend the elastic moduli of a rock is its hardness. A hard rock is difficult to compress because of high bulk (incompressibility) and high shear modulus (rigidity), and shows increase in P- and S-seismic velocities. Likewise, a soft rock with a large compliance has lower elastic modulii and consequently exhibits lower velocities. In a geological sense, elasticity may be likened to a measure of the hardness of a rock, which depends on lithology which commonly increases with depth.

Bulk Density

Bulk density of a sedimentary rock includes the density of the rock matrix and the density of the fluid in the pore spaces. Density of a rock is defined as its mass per unit volume and commonly increases with depth. It is a result of compaction, as the rock undergoes burial, the pore voids get compressed and the rock gets denser (Fig. 5). Compaction is a diagenesis process that squeezes out water from the pore space of sediments with time (depth) by

overburden pressure as they get buried beneath successive layers of sediments. Compact rocks show higher densities whereas under-compacted formations demonstrate lower density values. It may seem paradoxical that compact rocks at depth, though have higher bulk density, yet show higher velocities. This is because of relatively higher increase in elasticity of the compact rock than the increase in density with elasticity playing the dominant role in determining the velocity. It also may be stressed that velocity and bulk density are not directly related, though empirical equations exist which allow estimation of compressional velocities from bulk densities but limited to certain stipulated conditions, such as water-saturated and normally pressured sedimentary rocks (Gardener and Gregory 1974). Nonetheless, density determination from seismic remains a difficult task.

Porosity, Pore Size and Shapes (Pore Geometry)

Porosity is a measurement of the void space in a given volume of rock. In general, an increase in porosity lowers the density (Fig. 5) and more so the elasticity of a rock which results in decreasing the seismic velocity, more conspicuously the P-velocity. Though there is an established relation between porosity and density, no such

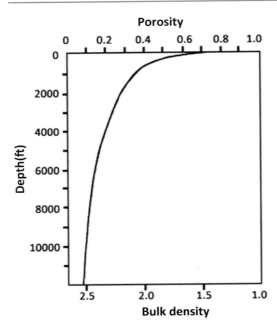

Fig. 5 Graph showing density increase with depth due to compaction of rock (diagenesis) in normal pressured sections. Compaction leads to reduction in pore voids, lowering the porosity resulting in increase indensity (after Anstey 1977)

definitive relation exists between porosity and velocity (Anstey 1977). Porosity and pore shapes are found to vary greatly in different kinds of rocks. Porosities are of two types, primary and secondary. Primary porosity is the pore space that occur between the grains in a sedimentary rock and are also known as intergranular porosity, as is seen in unconsolidated sands and partially cemented pores of sedimentary rocks. Secondary porosity, on the other hand, is pore space that has developed later on after the rock is formed. This may be caused due to fractures, weathering related leaching, solution channels and vugs linked to water ingression. Cemented sands and tight carbonates often exhibit void spaces because of fractures and cracks which are secondary porosities and are called fracture porosity. Similarly, the other types of secondary porosities depending on their causatives are known as leached, solution and vuggy (cavernous) porosities, mostly common in carbonate rocks. The type and order of porosity, controlled

by the pore geometry, determine the seismic properties, the velocity and density generally decreasing with increasing porosity.

Nevertheless, an empirical time-average equation (Wyllie et al. 1956) provides a basic link between primary porosity and velocity that is often used in interpretation.

$$1/V_r = 1 - \phi/V_m + \phi/V_f \, ,$$

where, V_r, V_m & V_l are velocities of the whole rock, rock matrix and liquid content in pore space, respectively and φ is porosity. It is known as time-average equation as the total time taken for a wave to travel in the rock is assumed to equal the sum of travel times in each rock component (Fig. 6). The time-average equation, however, has, several limitations such as particular types of rock with specific properties that include degree of porosity, type of fluid and normal pressure amongst others. It can be used to reasonably predict intergranular porosity of highly porous, water and brine saturated sandstones under normal pressure (Gregory 1977; Anstey 1977) but may not be applicable for highly porous gas saturated unconsolidated sands and in over-pressured regimes. The equation has since undergone several modifications (Raymer et al. 1980; Wang and Nur 1992) for present applications.

Pore geometry defines the distribution of voids and their size and shapes and is dependent on geometry of the grains of detrital sediments during deposition (Tatham 1982). The pore shapes can be of several types, ellipsoidal, spherical and penny-shaped, and are commonly described by the parameter, the aspect ratio which is the ratio of small to long axis of the pores. For flat penny shaped voids the aspect ratio is less than one and for spherical pores it is one. Rocks with flat pore shapes with small aspect ratio are amenable to higher compressibility and generally show lower velocity than those with large value aspect ratio such as with spherical pores. Pore shapes significantly impact behavior of P-velocity and sometimes more than the degree of porosity, per se. For instance, a reservoir with relatively lower porosity with flat

Fig. 6 Schematic illustrating the Willey's 'time average equation', linking velocity and porosity of a rock. The total time taken for a wave to travel in a porous rock is assumed to be equal to the sum of travel times in the two principal components, i.e. the matrix and the pore space filled with fluid

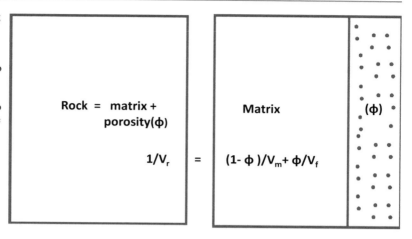

Rock = matrix + porosity(ϕ)

$$1/V_r = (1-\phi)/V_m + \phi/V_f$$

Matrix   (ϕ)

and low aspect - ratio pores may indicate lower seismic velocity than a highly porous reservoir with high aspect-ratio such as the spherical pores (Wang 2001). Though impact of pore geometry in velocity may be considerable, it remains a difficult parameter to quantify from seismic.

Texture

Grain sizes, roundness, sorting and cementation commonly describe the texture of a rock. Elasticity and density of a rock depend on contacts between the grains, their size and angularity though the latter ceases to play a role after the rock is cemented. Large grain sizes and compact sands have generally higher seismic properties due to larger contact areas causing higher velocity (elasticity) and density, whereas, unconsolidated sands with angular grains show lower seismic properties (Wang 2001)

Fractures and Cracks, Geometry

Seismic properties are affected considerably by presence of fractures (open) and cracks in rocks. Open fractures and cracks are considered different from void spaces like pores, caverns and vugs because their impact on elastic properties is disproportional to the much smaller volume of pore voids. Fracture porosities are commonly less than 2% (Gurevich et al. 2007) but can affect seismic velocity considerably. Fracture and crack geometry is more complicated than pore geometry; in addition to flatness, size and shapes of voids, it needs other parameters such as length,

orientation and density of the fracture and cracks and their distribution to define the fracture geometry fully.

Fractures and cracks usually facilitate compressibility (compliance) and considerably lower the velocity and impedance of rocks. Though fractures and cracks are known to impact significantly the seismic properties, it may be difficult to predict their geometry from seismic response. For instance, in a given volume of a fractured carbonate reservoir having a specific bulk density, similar fracture porosity can be expected either by assuming a large number of microfractures or by fewer numbers of bigger fractures. Even though the fracture porosity remains same, the velocity can be much lower in the former case compared to the latter (Anstey 1977). Sayers (2007) also indicated that the seismic response may be same for a small number of highly compliant fractures as for a large number of stiff fractures. This can have a significant implication on reservoir evaluation as the micro fractures linked to lower velocity may not be indicative of better permeability than the reservoir having larger fractures.

In case of cemented fractures, seismic velocity may indicate much higher values compared to what is expected at that depth. Such anomalous high velocities for a rock in a known tectonic area may corroborate the presence of fractures predicted from geologic data but it also offers a clue that the fractures are not open and may be cemented. Similar to pore shape geometry,

geometry of cracks and fractures with varying aspect ratio, too affect the seismic properties intricately and for flat-shaped fractures being significantly lower. The number and shape of fractures also determine the elasticity (compliance) of a rock which primarily decides the seismic properties, the velocity and amplitude. Another important aspect of fractures is causing anisotropy in a rock leading to azimuth dependent seismic properties discussed later. It also causes direction dependent wave attenuation and scattering linked to induced heterogeneity in the rock created by contrasts in elastic properties of open fractures with the surrounding rocks.

Anisotropy

A rock medium is considered anisotropic when the properties vary depending on direction of measurement. It is a vectorial variation of a physical property dependent on direction from a point and is different from heterogeneity which is a variation of property in scalar values limited to its position in a medium. Anisotropy induces heterogeneity in a rock but the reverse may not be always true; heterogeneity may not be a

necessary condition to create anisotropy in a rock. A simplified concept of the rock properties, heterogeneous, homogenous, isotropic and anisotropic is illustrated in Fig. 7.

Anisotropy in rocks is linked to stress and is commonly of two kinds, intrinsic and induced (Wang 2001).

- Intrinsic anisotropy is an inherent property of a rock caused by preferential alignment of grains and layering as in shale sedimentation. This is referred as VTI (vertical transverse isotropy), as it is associated with the vertical dominant stress, the gravity.
- Induced anisotropy in a rock, on the other hand, is caused by fractures and cracks and is referred as HTI (horizontal transverse isotropy), as it is associated with regional horizontal stress.

Seismic anisotropy is defined as direction dependent seismic velocity in a rock. Anisotropy affects both P-and S-wave velocities, though differently, being less perceptible in the former. In an intrinsic anisotropic medium (VTI) such as

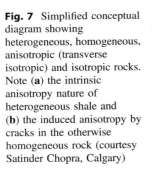

Fig. 7 Simplified conceptual diagram showing heterogeneous, homogeneous, anisotropic (transverse isotropic) and isotropic rocks. Note (**a**) the intrinsic anisotropy nature of heterogeneous shale and (**b**) the induced anisotropy by cracks in the otherwise homogeneous rock (courtesy Satinder Chopra, Calgary)

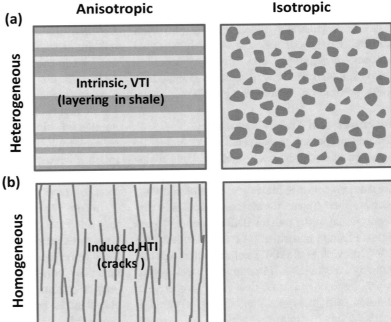

shale, the P-velocity is faster along the layered bedding planes than in the vertical direction (Fig. 8a). Similarly, in an induced anisotropy medium (HTI) such as in fractured rocks, P-and S-velocities travelling parallel to fracture plane are faster than those travelling across (Fig. 8b). However, variance of seismic velocity with azimuth, also termed as azimuthal anisotropy is more conspicuous in S-waves. The S-wave in HTI media splits to two waves travelling with different velocities in directions parallel and orthogonal to the fracture orientation. The wave splitting phenomenon is known as 'birefringence' and discussed in Chapter "Shear Wave Seismic, AVO and Vp/Vs Analysis". The direction dependent velocities in intrinsic anisotropic media also result in varying amplitudes with offset or angle of incidence, as evinced in wide-azimuth, far- offset seismic data.

Fluid Properties of Rocks

Pore Fluid and Saturation

Most sedimentary rocks have fluid in pore space. Fluids typically are known to have negligible shear modulus but affect compressional seismic properties depending on its compressibility and density. In a fully water or brine-saturated reservoir rock, water or brine offers resistance to stress and tends to increase velocity though not to the same extent as in a tight rock having little water. Oil saturation in rock pores lowers velocity marginally compared to that with water, as the comparatively lower bulk modulus of oil is offset to some extent by its lower density. However, based on velocity, it is usually hard to distinguish one from the other. In general, rocks saturated fully with liquids exhibit increased seismic properties (Wang 2001). Gas, on the other hand, has the least bulk modulus (highly compressible) and density, and the velocity and impedance of a rock with gas in the pore tend to show significantly lower values than that of rocks saturated with water and/or oil. The lowering of seismic velocity due to presence of gas, even in small quantity, is conspicuously large, especially at shallow depths (Fig. 9). Overall, the effect of fluid-saturation on seismic velocities decreases with increasing depth.

Seismic properties are influenced by both rock matrix and fluid properties and given the reality of wide variations that commonly occur in nature, it may not be always easy to differentiate the effect of one from the other. However, presence of gas in rock pores is more readily detectable as it considerably impacts seismic properties. But to

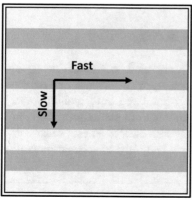

(a) VTI medium (shale)   **(b) HTI medium (fractured)**

Fig. 8 Schematic diagram illustrating seismic anisotropy in vertical and horizontal transverse isotropic media. Note the direction dependent velocities in VTI and HTI anisotropic media, (**a**) faster velocity along the bedding plane and (**b**) along the fracture orientation (courtesy S. Chopra, Calgary)

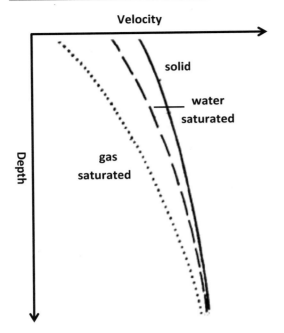

Fig. 9 Variation of velocity with depth for solid, water and gas saturated rocks at normal pressure. Velocity variation is significant at shallow depths but tends to be marginal at greater depths (after Anstey 1977)

estimate partial fluid saturations, that is, the percentage of water, oil and gas in rocks from analysis of seismic properties is difficult. It is interesting to note that though gas in most cases can be easily detected, estimating its saturation is a challenge. This is because as low as 5–10% saturation and 100% gas saturation is known to result in very similar seismic response.

Gassmann's Equations

Nevertheless, several physical and numerical methods have been deployed to model and study impact of fluid saturation on rock velocity. But by far the most widely used relations are the Gassmann's equations to calculate effect of different fluid saturations on seismic properties. It computes the impact of fluid saturation on bulk modulus in porous medium using the known bulk modulii of the other elements of a porous rock – the frame, the matrix, and the pore fluid by modelling (Wang 2001). There are, however, several assumption to Gassmann's equation,

including a major one that the porous material is isotropic, elastic, and homogeneous, which is often not the case in many hydrocarbon reservoirs. In fact, most reservoirs are known to be heterogeneous and mildly anisotropic. Particularly, for reservoirs having intricate pore and crack geometry with voids of varying aspect ratios, the Gassmann's equations do not apply well. Kuster and Toksöz (1974) developed a method for such situations where velocities are calculated taking into consideration the impacts due to different aspect ratios of pores. For a given porosity, seismic velocities increase as the pore aspect ratio increases

Permeability

Permeability is the property which denotes the ability of a fluid to flow in a rock. It has no linkage whatsoever with elasticity and density to influence seismic properties. High porosity often estimated may not be the effective porosity which is about the connectivity of the pores. Effective porosity computed from well logs, though is the closest, it is not the actual measure of permeability. Permeability depend on pore throats and tortuosity in a rock and can be measured from cores in laboratories. Unfortunately, permeability, the most important fluid property cannot be predicted from seismic.

Viscosity

Rocks tend to exhibit increasing elasticity and density with increase in viscosity of oil. Heavy oil has large bulk modulus and in some cases may tend to act as semisolids in the rock pores (Wang 2001). These rocks obviously exhibit relatively higher seismic properties.

Pressure

Besides the rock and fluid properties, pressure and temperature at depth also influence seismic response. Ignoring the horizontal tectonic stresses, rock at depth is basically under two vertical stresses, opposing each other. These are the overburden (geostatic or lithostatic) pressure, also known as the confining pressure and the fluid pressure or pore pressure (formation pressure). Overburden pressure at a depth is the pressure exerted by the overlying rocks acting downwards due to gravity while the pore pressure of fluid in the rock pores acts upwards due to buoyancy of fluid. While the overburden pressure tends to close the pores, the opposing pore pressure tries to retain the voids. The difference between these two pressures, is known as the effective pressure or differential pressure and is an important factor in influencing seismic properties. Change in effective pressure impacting closing or opening of pores and cracks, results in increase or decrease of elastic moduli of the rock. Higher effective pressure increases seismic properties and vice versa.

Normal Pressures (Hydrostatic Pressure)

As deposition continues, sediments get buried in depth and starts getting compacted under loading by expelling water. In properly compacted sections where the expelled water escapes to the surface, it maintains hydraulic communication with it and the formation shows hydrostatic pressure, commonly termed the normal pressure. Hydrostatic pressure, typically has a gradient of 0.43psi/ft; the pressure gradient defined as the rate of change in formation fluid pressure with depth. Hydrostatic pressure, however, is controlled by the density of the fluid saturating the formation. Fluid density in formation changes with depth due to temperature and pressure and also with the type of fluid. Consequently, brine saturated rocks show higher hydrostatic pressure gradient than the oil and gas saturated rocks exhibiting lower gradients.

In normally-pressured sections, effective pressure increases with depth because of increasing overburden pressure and raises elasticity and density of the rock resulting in higher seismic properties. The increase in velocity,

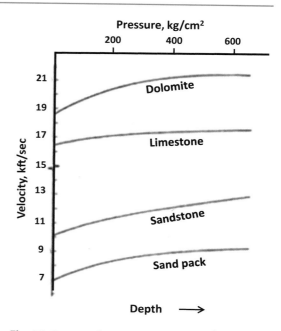

Fig. 10 Impact of pressure on seismic velocity for different rocks. P-velocity increases with increasing effective pressure, variation more conspicuous for rocks at shallow depths in lower pressure regimes (modified after Gregory 1977)

however, is nonlinear with depth and is more pronounced at shallower depths and in lower ranges of effective pressure (Fig. 10). The degree of change, however, varies with lithology, depending on elasticity of the rock (hardness), being maximum in soft unconsolidated sands and minimum in limestones.

Abnormal Pressures

Abnormal pressures are often labeled improperly as overpressures. Abnormal pressure is what is not normal and it can be higher or lower than normal pressures. Pressures higher than normal hydrostatic pressure are known as overpressured and lower as underpressured or subnormal-pressured.

Overpressures

Overpressures are generated by several mechanisms, the more common being the compaction or '*loading*' and fluid volume expansion or '*un-loading*'. In some geologic settings, formations undergoing subsidence release pore water under compaction, which, however, cannot escape to surface because of impermeable rocks at its top acting as seal, rocks such as shale or tight lime-stones. The expelled water thereby is forced to stay within the formation resulting in raising the pore pressure. This can happen in areas where rapid subsidence with huge amount of sediment loading occurs. If the rapid loading of sediments increases the overburden pressure at a rate which the escape of released fluid cannot keep pace with, the pore fluid has to support a large part of the load and thereby increasing the fluid pressure. The formation without going through normal compaction, remains undercompacted and exhi-bits fluid pressure higher than normal hydrostatic pressure and the formation is termed over-pressured (Fig. 11).

Overpressures due to the fluid volume expansion mechanism is known as '*unloading*'. This usually occurs in low permeability rocks such as shale, where fluid changes phase mainly due to generation of hydrocarbon during thermal cracking of organic matter, described later in Chapter "Seismic stratigraphy and Seismo-tectonics in Petroleum Exploration". Overpres-sures may also happen due to clay diagenesis and other causes such as by tectonics and geothermal heating where increased volume of fluid, con-strained by the limited rock matrix exerts higher fluid pressure (Fig. 12). The over-pressured rocks compared to a normally pressured layer, at the same depth characteristically show lower effective pressure, higher pore pressure, decreased elasticity, interval velocity and bulk density. The decreased trends of density and velocity typically tend to remain constant in the high-pressured zone despite increasing depth of burial (Figs. 13 and 14). Continued thickening of the overburden in such cases does not affect the seismic velocity in the zone as the high pore-fluid pressure continues to withstand the increasing part of the overburden pressure with depth. Overpressures and pore pressures can be detected from sonic and density logs by their character-istic anomalous features of density remaining

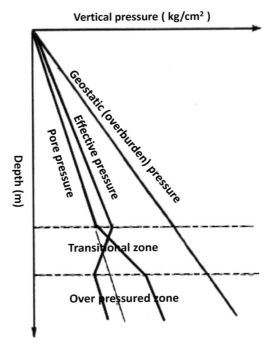

Fig. 11 Schematic showing geostatic, hydrostatic (nor-mal) and effective pressure (difference between geostatic and pore pressure). In under-compacted rocks, the unexpelled water remaining in pores cause higher pore pressure than hydrostatic and are known as over-pressured rocks. Over-pressured formations show reduced effective pressure

constant and velocity reversing trend with depth despite increasing depth (Fig. 15). Overpressures can also be indicated from seismic velocity, discussed in Chapter "Evaluation of High-Resolution 3D and 4D Seismic Data".

Underpressures or Subnormal Pressures

The causes of subnormally low pressured forma-tions are not clearly understood. However, when a saturated porous formation is subjected to uplift and erosion, the overburden pressure is reduced causing a partial elastic rebound of the reservoir matrix which was previously under compaction. The increase in volume of the pores eases pore pressure to reduce pressure and the formation exhibits subnormal pressures. Compaction in shales due to increase in depth is inelastic in

Fig. 12 Schematic illustrating loading and unloading mechanisms of overpressure formations with the lines and curve indicating the related stresses—the geostatic (overburden), hydrostatic (normal), overpressures and the effective pressure (after Chopra and Huffman 2006)

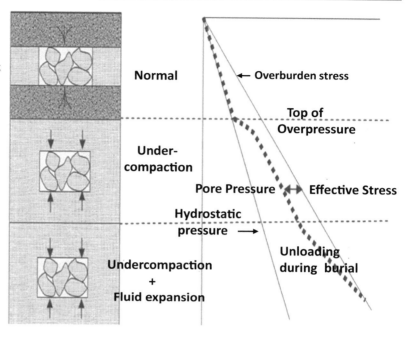

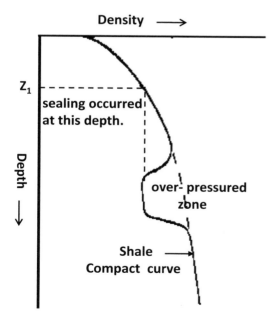

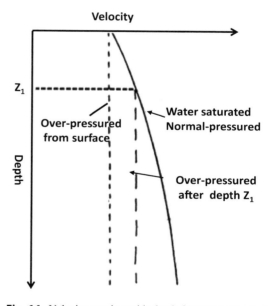

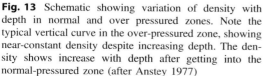

Fig. 13 Schematic showing variation of density with depth in normal and over pressured zones. Note the typical vertical curve in the over-pressured zone, showing near-constant density despite increasing depth. The density shows increase with depth after getting into the normal-pressured zone (after Anstey 1977)

Fig. 14 Velocity varying with depth for water saturated rock under normal pressures (compaction curve) and over pressures. Over-pressured zones show lower velocity which tends to remain constant despite increase in depth. The point where velocity drops (z_1), indicates the depth from where over pressures started (after Anstey 1977)

Fig. 15 Example of typical (**a**) density and (**b**) sonic velocity logs showing characteristics of normal, undercompacted and overpressured sections. The trend line at top signifies normal compaction, in the middle undercompaction and at the bottom, the overpressured formations. Note the constant density and the velocity reversal trends in bottom part of (**a**) and (**b**) curves, characteristics of overpressured zones (after Chopra and Huffman 2006)

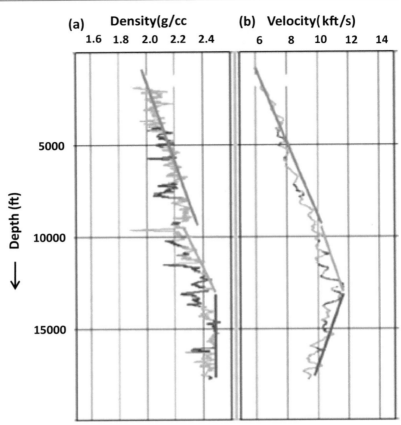

nature whereas, highly porous formations such as sands it is elastic making it more amenable to recover to original state though not fully. Evidence of most low pressure reservoirs associated with uplift and erosion happening to be sandstones, provide as corroborative evidence.

Temperature

Temperature influences fluid properties such as viscosity and elasticity. Increase in temperature, decreases seismic properties marginally in water and gas saturated rocks but decreases significantly in rocks saturated with heavy-oil. Increasing temperatures by heating heavy-oil, especially in unconsolidated sands (e.g. tar sands), can result in a remarkable decrease of viscosity and enhance oil mobility. This impacts seismic properties as mobility of oil promotes compressibility of the rock. Heating at high

temperature may also generate new cracks in the rock and thereby decrease the elastic modulii of the rock. Thermal methods used to heat heavy-oil for thinning to flow are common for recovering heavy oil from sand reservoirs (discussed in Chapter "Oil Sands, Heavy Oil, Tight Oil/Gas, BCGAs and CBM").

Seismic Rock Physics Modelling (RPM)

Important rock-fluid properties of a reservoir such as lithology, porosity, permeability, fractures, texture, fluid type and saturation etc., are known at a well. But almost all reservoirs are heterogeneous with the properties varying laterally and vertically within the reservoir. How does one estimate and predict these changes in rock parameters, away from the well? Seismic rock physics modelling enables to provide benchmarking of the known rock parameters at the well with seismic response. Once this is accomplished, seismic data can be interpreted and used to predict the properties over the entire prospect. Modelling is thus used as an effective mean to build a predictive model to estimate reservoir parameters. Rock physics modeling is an important tool providing the link between reservoir and seismic properties that allows quantifying reservoir parameters at places between and away from the wells and seismic rock physics modelling (RPM) has since, been an integral part of seismic interpretation work flow.

While estimates of rock properties, such as, lithology and porosity from seismic data have long been a part of routine interpretation, quantitative estimate of the subtler fluid properties like saturation and permeability remains a big challenge. This is due to many intrinsic limitations in seismic data in its manifestation and interpretation and also perhaps due to lesser understanding of rock physics and its application. For example, it is long since reported that some oil/gas saturated reservoirs show low frequency shadow zones below a gas reservoir supposedly due to high energy absorption in gas, compared to water saturated rocks. This is an explanation which remains contentious. According to Ebrom (2004) who has extensively researched the phenomena, a convincing good explanation for the observed low-frequency is yet to come up. Ebrom in his paper has suggested several other possibilities that could be responsible for the phenomena (Refer Chapter "Direct Hydrocarbon Indicators (DHI)"). Another issue not too well understood and under controversy is the effect of 'Fizz gas' on seismic properties. 'Fizz gas', refers to noncommercial or residual amount of gas such as dissolved gas in water or small amount of free gas.

Mechanical Properties of Rocks, Geomechanics

Mechanical properties of rocks (rock mechanics) involve their behavioral changes under applied stress, pressure and temperature and study of the branch is known as geomechanics. Though geomechanical applications are commonly limited to civil & mining engineering, it is becoming increasingly important in petroleum industry, as well, in reservoir characterization, reservoir simulation and stimulation, production and drilling engineering applications. This is known as 'reservoir geomechanics' and is briefly outlined in Chapter "Evaluation of High-Resolution 3D and 4D Seismic Data" under 4D seismic applications. Geomechanical properties analyzed in hydrocarbon exploration for monitoring reservoir stimulation for secondary recovery and in drilling applications for proper well completion are primarily the Young's modulus (E), the Poisson's ratio (σ), pore pressure and the in situ stress which are estimated from seismic.

Young's Modulus

Young's modulus, besides bulk and shear modulus, is another primary elastic modulus defined as the stress-strain ratio. It denotes the ability of a rock to stretch (deform) lengthwise under extensional stress. The strains are related to the applied stresses by the rock elastic modulii, which include bulk modulus, shear modulus, Young's modulus, and Poisson's ratio. Young's modulus measures the stiffness of rock, also referred as 'stiffness moduli' and the shear modulus the rigidity. Young's modulus and shear modulus are related through impedances by the equation

$$E\rho = \mu\rho\left(3Pi^2 - 4Si^2\right)/\left(Pi^2 - Si^2\right)$$

where, Pi and Si are the P-and S-impedances and ρ the density. The equation also shows. $E\rho$ is a scaled version of $\mu\rho$ (Sharma and Chopra 2015).

Estimation of Young's modulus is extremely important to determine stiffness in rocks for fracturing, carried out in shale and tight reservoirs for hydrocarbon production (Chapter "Shale Oil and Gas, Oil Shale and Gas Hydrates"). The equation helps calculate seismic attributes, products of Young's and shear modulus and density to determine shale brittleness, a requisite rock property, necessary for fracturing. The modulus and density product attributes have the advantage that it does not require the density value which is usually difficult to determine from seismic. The product of shear modulus and density ($\mu\rho$) is also considered indicator of rock stiffness but product of Young's modulus and density ($E\rho$) is more sensitive (Fig. 16) to be preferred. Generally the Young's modulus is considered as the indicator of rock brittleness and the product attribute ($E\rho$) with high values denote high rock brittleness (Sharma and Chopra 2015).

Poisson's Ratio

Poisson's ratio ('σ)' is an elastic constant defined by the ratio of transverse strain to longitudinal strain which corresponds to the ratio of bulk modulus to shear modulus, the 'k/μ' (Fig. 17).

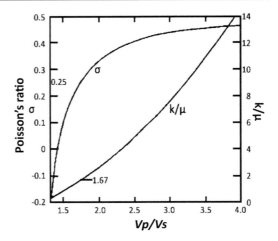

Fig. 17 Graph showing the inter-relation of Poisson's ratio (σ), bulk to shear modulus ratio (k/μ) and velocity ratio (Vp/Vs). Increase in one shows increase in other ratios as all the ratios vary similarly though not linearly (after Tatham 1982)

Simply expressed, lower the bulk modulus (k) of a sedimentary porous rock, smaller is the likely value of 'σ'. Poisson's ratio is often expressed for convenience by the ratio of P and S velocities, the Vp/Vs of a rock and is related to Poisson's ratio by the equation,

$$\left(\frac{V_p}{V_s}\right)^2 = \frac{2(1-\sigma)}{1-2\sigma},$$

From the equation it can be seen that both vary in the same way though not linearly; higher Poisson's ratio corresponds to higher Vp/Vs values and *vice versa* (Fig. 18).

Poison's ratio represents the strength of a rock. It may be stressed that strength and stiffness of a rock are different geomechanical properties; strength is a measure of stress that the rock can handle before it deforms permanently, whereas stiffness is a measure of resistance to elastic deformation. Young's modulus and Poisson's ratio can be determined from sonic and density logs or from seismic data (Chapter "Seismic Modelling and Inversion"). More about Poisson's ratio, significance and application in estimating reservoir parameters is discussed in Chapter "Shear Wave Seismic, AVO and Vp/Vs Analysis".

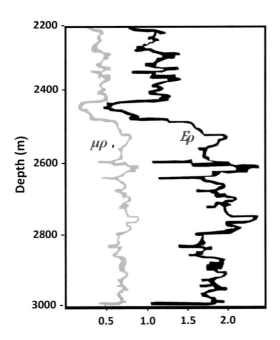

Fig. 16 Product attributes of density with Young's modulus ($E\rho$) and shear modulus ($\mu\rho$). Note the higher sensitivity of ($E\rho$) attribute compared to ($\mu\rho$). Higher values indicate more brittleness of rock sensitive (after Sharma and Chopra 2015)

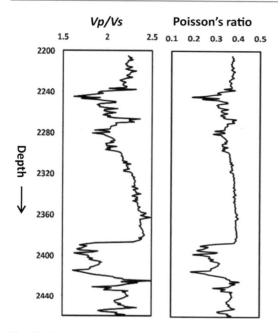

Fig. 18 Example showing log computed *Vp/Vs* and Poisson's ratio (σ) curves. Note the excellent correspondence between the two curves, changes in one reflect similar changes in the other (courtesy ONGC, India)

In Situ Stress

In situ stress is a stress confined in a rock formation and consists of three mutually orthogonal stresses, the vertical stress and the two principal horizontal stresses, the maximum and the minimum. However, sedimentary rocks contain fluids in their pore spaces and exerts also pore pressure opposing the overburden force and is also taken into consideration.

Mechanical Earth Modelling (MEM)

Mechanical properties of rock vary widely, laterally and vertically depending on many factors, the most important being the lithology of a rock and its composition, overburden and pore pressure, minimum and maximum principal horizontal tectonic stress, temperature, and depth of burial. In the context of petroleum exploration, it is essential to assess the geomechanical properties of rocks to judge how rocks would deform or

change behavior in response to drilling, reservoir stimulation and fluid production to ensure trouble-free maneuvers. This necessitates understanding mechanical properties of rocks, similar to physical properties, to predict the behavioral changes so as to conduct safe and fast operations. Mechanical earth modelling (MEM) is mostly based on the three parameters the Young's modulus, the Poisson's ratio and in situ stress discussed above. The modelling can be one dimensional around the well bore analogous to synthetic seismogram or three dimensional in a field or regional scale.

Mechanical earth modelling has important applications for drilling engineers. Well bore instability causing whole collapse, lost mud circulation and stuck pipe are some of the problems faced during drilling that lead to increased operational cost and time. Prior knowledge of the pressure and rock parameters can help drilling engineers plan, design and optimize parameters such as mud weight and casing programs to avoid troubles for safe and economic drilling and timely well completions. High pressured zones, which can be hazardous also need caution for drilling engineers. Geomechanical effect concerning field production is another field of application is dealt under reservoir geomechanics. (Chapter "Evaluation of High-Resolution 3D and 4D Seismic Data"). Geomechanics impacting process of stimulating the reservoir for production are discussed in Chapters "Oil Sands, Heavy oil, Tight Oil/Gas, BCGAs and CBM and Shale Oil and Gas, Oil Shale and Gas Hydrates".

Mechanical earth models are usually made by estimating mechanical rock properties using wireline log, drilling, and core measurements. As regards insitu stress, the overburden and pore pressure which are linked to density and velocity of rock can be calculated from density, sonic and resistivity logs. However, pressure measurements in the well, if done, directly provide the pressure and pressure gradient. Of the two principal horizontal stress, only the minimum horizontal stress is critical as its strength and orientation controls the magnitude and orientation of the deformation in the rock. This can be obtained from the leak off tests (LOT) conducted during drilling. Leak off tests are carried out to indicate the maximum pressure or mud-weight that can be applied to the formation. Minimum horizontal stress direction can also be determined from image logs and four arm caliper log which measures the borehole size.

One dimensional mechanical earth model can thus be generated around the well, but can have prediction risks away from wells. Often "geostatistical tool" such as krigging, co-krigging are used to interpolate the data between wells if available. However, a dependable model can be made using high resolution 3D data which can generate the elastic, pore pressure and stress parameters to fill in the gap between the wells and beyond. However, this would require stringent calibration with well log, leak off tests, and pressure measurement data in the well.

Limitations of Modelling

Most rock physics studies about rock-fluid properties are either from measurement of samples in laboratories or based on empirical observations from geologic specific areas. Often the properties determined are computed from numerical modelling. These can have mismatches with those determined from real field seismic data due to several inherent reasons. The lab-measured samples are not true representative of insitu samples where the conditions are completely different. One of the major limitation is the critical factor of dimension and the adjusted scaled factor leading to discrepancies in one-to-one correspondence of laboratory or log measured rock-fluid properties with seismic. While measurements in logs and laboratories are on a microscopic scale (mm to cm) the seismic measurement is in macroscopic scale (tens of metres). Though often a scaling factor is used as a transform factor for up or down scaling, inconsistencies may still remain between the sets of estimated values. Seismic scale of measurement has another concern when stretches of heterogeneity in a rock, could be imaged by seismic as homogeneous and isotropic due to its limited bandwidth. A further problem which can be

critical for disparities is that while the mechanical properties of rocks measured in laboratory provide static modulii, the elastic wave velocity and density determined from seismic express dynamic moduli of rocks.

The sensitivity of seismic properties to changes, especially in fluid properties of a rock, is subtle and its detection and interpretation in a real situation remains a formidable and challenging task for the analyst. Modelling does mitigate problems to a large extent but uncertainty still remains in seismic predictions of rock fluid properties. Different constitutional elements of a rock may act against one another thereby reducing their impact on seismic response to be perceptible. Nevertheless, in some favorable geological and petrophysical setups, such as in highly compliant, shallow unconsolidated gas saturated sands, the effects of individual rock-fluid parameters may be additive so as to result in a seismic response that can be clearly discernible.

It may also be noted that the aforesaid discussions are all about compressional seismic properties involving only P-waves. Some of the rock-fluid parameters are known to be differently sensitive to shear waves and a combined analysis of properties of P and S waves is often found to be much more useful. This is dealt in Chapter "Shear Wave Seismic, AVO and Vp/Vs Analysis" under the topic of shear seismic.

References

Anstey AN (1977) Seismic Interpretation, The physical aspects, record of short course "The New Seismic Interpreter". IHRDC

Chopra S, Huffman Alan (2006) Velocity determination for pore pressure prediction. CSEG Recorder 31:1–29

Gurevich B, Galvin RJ, Brajanovski M, Muller TM, Lambert G (2007) Fluid substitution, dispersion, and attenuation in fractures and porous reservoirs- insights from new rock physics model. Lead Edge 1162–1168

Ebrom Dan (2004) The low-frequency gas shadow on seismic sections. Lead Edge 23:772

Gardener LW, Greogory AR (1974) Formation velocity and density-the diagnostic basics of stratigraphic traps. Geophysics 39:770–780

Gregory AR (1977) Aspects of rock physics from laboratory and log data that are important to seismic interpretation. AAPG Memoir 26:15–46

Kuster GT, Toksöz MN (1974) Velocity and attenuation of seismic waves in two phase media. Geophysics 39:587–606

Raymer LL, Hunt ER, Gardner JS (1980) An improved sonic transit time-to-porosity transform. In: SPWLA 21, annual logging symposium, pp 1–12

Sayers CM (2007) Introduction to this special section: fractures. Lead Edge 26:1102–1105

Sharma RK, Chopra S (2015) Determination of lithology and brittleness of rocks with a new attribute. Interpreter's corner. Lead Edge 936–941

Tatham RH (1982) The discrimination of fluid content and lithology in a reservoir. Vp/Vs Lithology Geophys 47:336–344

Wang Z (2001) Y2K tutorial-fundamentals of seismic rock physics. Geophysics 66:398–412

Wang Z, Nur A (1992) Seismic and acoustic velocities in reservoir rocks: geophysics reprint series 10, vol 2

Wyllie MRJ, Gregory AR, Gardener LW (1956) Elastic wave velocities in heterogeneous and porous media. Geophysics 21:41–70

Seismic Reflection Principles—Basics

Abstract

Seismic reflection events are caused by impedance contrasts at interface of two layers and with a requisite minimum width (Fresnel Zone). The reflection images are required to be of utmost quality to represent reliably the subsurface geology. The quality of an image depends on signal-to-noise ratio and its resolving power, the ability to show details of individual geologic features, vertically and laterally. This requires an efficient energy source to generate seismic waves with broad-bandwidth, consisting of both low and high frequencies that can penetrate deeper to detect and define stratal layers and their properties.

A seismic reflection trace provides attributes such as amplitude, phase, polarity, arrival time and a record comprising of multitraces the velocity which can be measured or estimated. The polarity information is important for calibration of seismic with well and for identifying lithologies and correlating strata. Reflection polarity is crucial particularly for authenticating high amplitude seismic anomalies for direct detection of hydrocarbon and its role is emphasized in the chapter along with the other attributes. The reflected wave-form and its arrival time depends on rock properties of a strata and its depth and carries these crucial geologic information, contained in the seismic attributes of the waveform. Estimating the rock properties from seismic waveforms and their vertical and lateral changes in time and space is the essence of seismic interpretation.

Display of data plays a major role in visualizing seismic images and appropriate choice of seismic display modes and plotting scales are important for optimal geologic perception of seismic image and is stressed with example.

Seismic wave, generated artificially on surface when propagates downwards through the earth, it meets several interfaces between different kinds of rocks having diverse properties. This causes phenomena such as reflections, refractions, scattering and diffraction. Of these, the reflection is by far the most significant phenomenon, as it forms the basis of the potent seismic reflection technology, deployed to portray the subsurface to find hydrocarbons. A compressional wave (P-wave) when incident normal to an interface, causes reflection and transmission also normal to it, but when incident at an angle (inclined), it produces two sets of waves, P-reflected and P-transmitted (refracted) and S-reflected and S-transmitted (Fig. 1). We shall limit the discussions for the present to the principles of relatively simple P-wave reflections, used extensively for measuring rock and fluid properties. The S-waves reflections, their properties and utilities are discussed in Chapter "Shear Wave Seismic, AVO and Vp/Vs Analysis".

N. C. Nanda, *Seismic Data Interpretation and Evaluation for Hydrocarbon Exploration and Production*, Advances in Oil and Gas Exploration & Production, https://doi.org/10.1007/978-3-030-75301-6_2

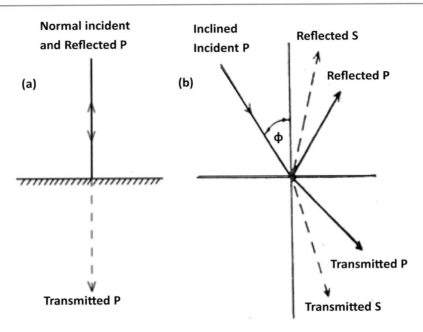

Fig. 1 Seismic wave propagation phenomenon at an interface of two different rocks. (a) normal incident P-wave produces single set of waves normal to the interface, the P-reflected and the P-transmitted, whereas

(b) inclined incident produces two sets of waves, P-reflected and P-transmitted and S-reflected and S-transmitted (refracted)

Seismic reflection event herein is considered as a correlatable event over an area conveying a geologic feature. Generation of a seismic reflection event needs primarily two things, (a) an impedance (product of velocity and density) contrast at the interface of two rock types and (b) a minimum width (Fresnel Zone) of the interface. The reflection amplitude and its continuity depend on the degree of contrast across the interface, its type and geometry and lateral extent. Effectiveness of reflection events to reliably portray the subsurface geology is, however, conditional to the quality of seismic reflection signal, which is influenced by (i) the amount of noise recorded in the data and (ii) the ability of the seismic wavelet to image the different interfaces separately and distinctly. The seismic reflection quality is thus adjudged by the two seminal factors, the signal-to-noise ratio(S/N) and the resolving power of the seismic wavelet, briefly described below.

Signal-to-Noise Ratio (S/N)

Noise may be defined as all types of undesired energy, other than the primary reflections from the subsurface strata that are recorded. Noise is an inherent part of the seismic data acquisition and processing system, created by intrinsic ambient (within earth) noise, geological (natural propagation hurdles in the subsurface) and geophysical (artifacts during recording and processing) processes. Noise cannot be wished away, but can be effectively reduced by conscious efforts during data acquisition and processing. Ironically though, noise usually considered unwanted, can be occasionally helpful in interpretation in some cases. For example, remnant diffraction noises despite processing may indicate clues to presence of sharp edges, such as faults and other subtle stratigraphic objects, particularly in 2D data. Scattering is another kind of noise which may give an idea about the order of heterogeneity of the reflector, leading to indication of highly tectonized zones with complicated structures and associated faults and fractures.

Since noise severely affects seismic clarity in portraying the subsurface image, it is desirable to record good and clean signals with minimum noise. It is a common practice to benchmark the quality of data in terms of a measure of a ratio between signals and noise (S/N). Improved data acquisition requires meticulous planning of survey layout plans, field experimentations and optimal parametrization followed by strict on-field execution to ensure data quality with good S/N ratio. The unique common depth point (CDP) technique for seismic data acquisition in the context is by far the most valuable gift to reflection seismic technology. It is a standard technique practiced all over the world with many possible variations in lay outs, according to suitability of the geologic objectives. It achieves signal enhancement at the cost of noise, via a summation process of several traces reflected from the near-same common depth point with different offsets after correcting for the offset geometry, known as CDP fold stack. In reality it is not exactly the same depth point because of the dipping layers in the subsurface and is more precisely termed as common md point (CMP) stack. What the summation does is it amplifies the signals which are similar and cancels the noise which are random. Though summation of higher number of traces in a fixed offset range generally provides better S/N ratio, there may be a limit beyond which it may not be desirable as adding more traces (folds) cost extra money for acquisition without commensurate improvement in the seismic images. Also summation is an integration process, which affects resolution due to loss of high frequency, especially in cases where large far offset traces are included for summing. In areas where the geology promotes good quality seismic reflections, the interpreter may still prefer to look at less-fold CDP data which is likely to offer better resolution and at a lower cost. It may also be noted that data with high S/N ratio does not necessarily assure higher resolution, as the resolution depends on

other factors such as source signal frequency, sampling interval and subsurface wave propagation effects, besides noise.

Seismic Resolution

Resolution may be defined as the ability to separate two closely spaced features in depth (time) as well as in space. The resolution we refer to in seismic is of two types, vertical and horizontal. Vertical (temporal) resolution is the minimum separation in time between two reflections arriving at the surface for recording that enables to detect each reflector separately. Lateral (or spatial) resolution is the minimum lateral distance between two closely spaced geologic objects in space that permits each one to be imaged separately. It is important to keep in mind that detection of an event is not the same as resolution which defines the detected object clearly.

Resolution depends on the seismic wavelength with which the subsurface feature is measured. Wavelength is a fundamental property of a wave which is the distance between successive points of its equal phase (e.g., crest to crest), completing one cycle. It is usually denoted by the symbol λ and is defined by the equation $\lambda = v/n$, where 'v' and 'n' stand for the velocity and frequency of the wave passing through a medium. Smaller wavelengths provide better resolution whereas wave lengths, too large compared to the dimensions of the object, fail even to detect it. Since wavelength is a direct function of velocity and inverse of frequency, seismic resolution happens to be better at shallow depths where higher frequencies are dominant and the seismic wavelength is smaller due to relatively lower velocity. On the other hand, because of increasing velocity and lowering of frequency with depths, the seismic resolution deteriorates with depth.

Vertical Resolution

A short-width sharp zero phase wavelet (high frequency bandwidth) ideally provides the best resolution. Zero phase wavelet is symmetrical with maximum amplitude at time zero, chosen as the origin and has small and even side lobes (Fig. 2). Because of short duration (in time) and the nature of the wavelet, reflection arrival times correspond to exact depth of the geologic features without time delay and thus facilitates beds to be imaged individually without overlaps and at appropriate time with respect to depth. In contrast, minimum phase wavelets are asymmetrical, front-loaded energy wavelet with uneven side lobes which impede resolution (Fig. 2). A zero phase wavelet is therefore an interpreter's desired wavelet though the commonly used seismic sources like dynamite on land and air-guns in marine surveys produce minimum phase wavelets. However, Vibroseis source used on land, generates a zero-phase wavelet, known as *Klauder* wavelet that makes it a preferred choice. However, Vibroseis trucks are not accessible in many terrains and also provides relatively less energy compared to dynamite source. More about the zero and minimum phase wavelets and their nature is described under subhead polarity.

The seismic short source wavelet, an impulse, however, while traveling within the earth suffers loss of high frequencies due to absorption and gets changed with passage of time to a long and cyclic ('leggy') wavelet which becomes a mixed phase wavelet. The large and leggy nature of the wavelet does not permit enough separation between the arrival times of reflections coming from closely spaced beds. This results in overlapping of the individual events and losing the ability to resolve the beds separately (Fig. 3). It has been demonstrated by Widess (1973) by modelling a wedge that $\lambda/8$ is generally the limit of bed thickness as the vertical resolution, below which thinner beds cannot be seen as resolved. Widess's wedge model envisages impedance contrasts as same at the top and bottom of the wedge and with the signage reversed (Fig. 4). However, in many geologic situations the impedance contrasts at top and bottom are likely to be different in values and are also of same signage as top and bottom, in which case, the Widess model of thin bed resolution limit may be different. Nonetheless, practical experience shows that in real-earth situations, where some amount of noise is always present in the data, $\lambda/4$ may be considered a reasonable wavelength as the vertical resolution limit for resolving beds. Vertical (temporal) resolution worsens with depth and generally varies from 10 to 15 m at shallow depths to 20–30 m at greater depths.

Exploration objectives (reservoirs) are often thin and require improved vertical resolution for proper delineation. Resolution can be enhanced during acquisition by deploying a broad-band wavelet as a source (dynamite) and by recording

(a)

(b)

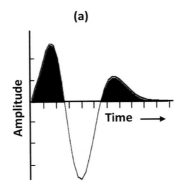
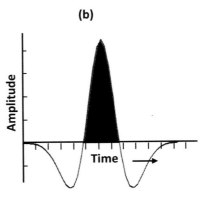

Fig. 2 Showing types of seismic source wavelet types (**a**) minimum phase wavelet, generated by dynamite on land and air-gun in offshore and (**b**) zero phase wavelet generated by Vibroseis in land data acquisition. Zero phase wavelet is symmetrical with even side lobes and the amplitude maxima occurs at time corresponding to exact depth of the strata without time delay. Note the minimum phase wavelet (**a**) is asymmetrical with energy loaded in front and the maxima occurs at a delayed time

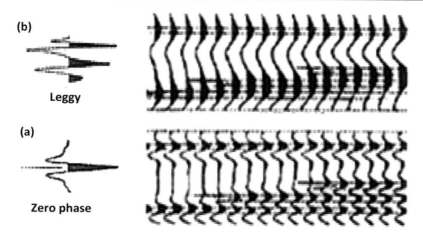

Fig. 3 Example showing resolving power of (**a**) zero phase wavelet and (**b**) leggy (cyclic) wavelet. Note the excellent vertical resolution of thin beds by zero-phase wavelet whereas, (**b**) the leggy mixed phase wavelet impedes resolution by causing overlapping of reflections from thin beds. Zero phase, short wavelet, shows maximum amplitude at zero time without delay and with small side lobes promotes better resolution (after Vail et al. 1977)

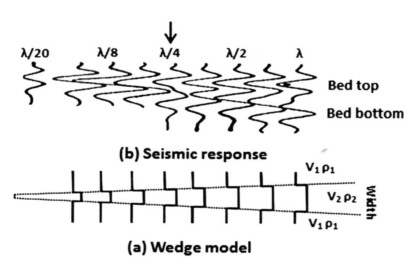

Fig. 4 Schematic illustrating Widess wedge model for vertical resolution limits. (**a**) The Widess wedge model and (**b**) seismic response for varying bed thickness. For a bed thickness greater or equal to the wave length (λ), the top and bottom reflections are clearly resolvable and are so till quarter wave length (λ/4). For beds thinner than λ/4 the top and bottom reflections are not distinct (arrow marked), limiting the vertical resolution to quarter wave length (after Widess 1973)

with smaller sample intervals (temporal, ∼ 2 ms). In addition to the data acquisition efforts, care is taken to retrieve the signals and improve resolution by boosting the higher frequencies during data processing, a technique known as deconvolution. The recorded seismic trace is a convolution, a mathematical process of conjoining two signals (likened to ∼ product), of the source wavelet with the earth's reflectivity series (array of impedance contrasts in the subsurface). If the source wavelet can be removed from the recorded trace through data processing, the impedance contrasts representing geologic rock discontinuities will be left behind which is the sole aim of seismic investigation. Deconvolution and zero-phase wavelet are processing steps to increase vertical resolution by suppressing multiples and by compression of the wavelet (shortening) that is achieved by increasing effective bandwidth and eliminating the effects of side lobes.

Lateral (Spatial) Resolution and Fresnel Zone, Migration

Huygens' Principle stipulates that reflection from a surface consists of a number of diffractions occurring from each point on it and does not come from a single point. Where the reflecting surface is uniform and planar, the diffractions from all points add constructively to provide a reflection event. If, however, the surface is curved or has limited small continuity, the diffractions may not add effectively resulting in poor reflection. Seismic waves that originate from a point source are spherical in nature, and when incident on a plane reflector, they sweep through it by producing a succession of contact zones. Nonetheless, the limited planar area, which 'effectively' comes into contact at the interface and collectively contributes to produce a coherent reflection event is called the (first) Fresnel zone (Fig. 5a). The seismic wave is a band-limited signal comprising a range of frequencies, and when incident on an interface, each frequency creates its individual area of contact with the interface to cause a reflection. However, the reflections recorded are considered to be from the first Fresnel zone formed by the dominant frequency. Fresnel zone is considered to serve as a yard stick for defining the lateral resolution, smaller the width of zone, better is the resolution. It is important to visualize the phenomenon of reflection as an 'area' concept in two dimensions and as 'volume' concept in three dimensions instead of a single 'point' notion that can have enormous significance in data interpretation and evaluation.

Reflectivity of widths less than a Fresnel zone tend to deteriorate the reflection quality.

Modeling has demonstrated that interfaces having width less than $\lambda/4$ cannot be viewed clearly and thus defines the limit for spatial resolution (Fig. 5b). The Fresnel zone may be considered as a spatial requirement, complimentary to the vertical impedance contrast, responsible for causing reflection event. Similar to thickness of the bed which determines the temporal resolution, the Fresnel zone width can be considered to serve as a yard stick for defining the lateral resolution.

The quality of a reflection depends not only on the area defined by Fresnel zone but also on the geometry and type of its reflecting surface. Warped surfaces behave as curved reflectors and depending on whether the reflector is concave or convex upwards, it exhibits focusing and defocusing effects. Consequently tight synclines are well imaged due to convergence of reflected ray paths while tight anticlines are poorly imaged because of divergence of rays (Fig. 6a, b). In extreme cases, acute synclines at greater depths with very high curvatures generate reflections converging from the concave reflector and cross one another before being recorded at the surface. This is known as a 'buried focus' effect and exhibits a familiar reflection pattern known as "bow-tie" (Fig. 6c). The bow tie apparently looks like antiform which in reality is a tight synform. Rougosity of reflectors such as erosional unconformity and highly heterogeneous and anisotropic rocks are other factors that cause large amount of scatter and impede image quality. Scattering is considered a noise and is usually not handled in routine data processing though recent advances in migration techniques can process efficiently the scatters in 3D data by reconstructing the energy from where it originated from.

Poor/no reflections at times, associated with fault edges, sharp facies changes, small reefal mounds and erosional unconformities may be examples of poor imaging linked to inadequate Fresnel's zone width. However, a small discontinuity in the reflecting surface, for instance, a hole cut in the Fresnel zone, will hardly affect the quality of the averaged reflection due to phenomenon known as 'wave front healing', a process by which the waves are diffracted around the

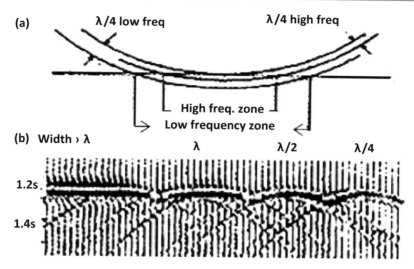

Fig. 5 Schematic illustrating phenomenon of reflection and the Fresnel's zone. (**a**) Spherical wave front incident on plane surface forms contact zones of different widths for each of the frequencies in the bandwidth. However, the contact area for the dominant frequency mainly influences creating reflection events and is known as the (first) Fresnel zone and is a measure of lateral resolution. Smaller the zone-width better is the resolution. The width of Fresnel zone at a depth is dependent on frequency, being larger for low frequency than for high frequency. (**b**) Synthetic reflection events computed with variable source wavelength. Notice the start of deterioration in reflection at $\lambda/4$, which sets this as the limit of spatial resolution (after Meckel and Nath 1977)

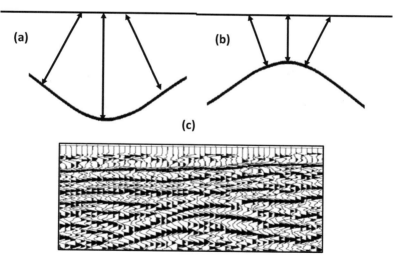

Fig. 6 Effects of curved surfaces on reflection. (**a**) rays reflected from concave reflector (synform) converge while (**b**) they diverge from convex reflector (antiform). Note the rays converging would cross one another before reaching the ground, if it is too far up. (**c**) This can happen for tight synforms at large depth and the effect is known as 'buried focus' and is manifested in unmigrated seismic as a 'bow tie', an artefact showing spurious antiform

anomaly (Fig. 7). This can have important geological implications, the open fractures and cracks present in rocks may be difficult to be imaged directly by seismic. Furthermore, Fresnel

zones in the subsurface are often not planar but consist of curved surfaces, which is yet another factor that affects quality of reflections particularly in 2D data. In perspective of Fresnel's zone

Seismic response

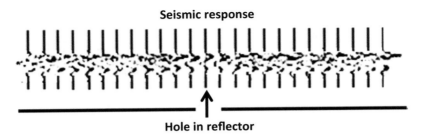

Hole in reflector

Fig. 7 Sketch illustrating 'wave front healing' effect. A small discontinuity in the Fresnel zone, in the shape of a small hole in the reflector, has little effect on seismic reflection because the diffracted waves go around it, and is known as 'wave front healing' (After Sheriff 1977)

width, for anticlines the contact area of the wave with the reflector is small amounting to lower amplitudes, whereas for concave surfaces (synclines), the contact area being more, provides strong amplitudes. This phenomenon is similar to focusing and defocusing effects of an optical lens as stated earlier. However, most of the effects discussed such as bow-tie, imaging of curved surfaces, focusing and defocusing and diffraction noise are usually not evinced in present day data as they are removed by powerful migration techniques during data processing. Nonetheless, interpreters may be mindful of these artefacts as in many cases 2D data of old vintage are only available for interpretation. Furthermore, hints of such features on seismic can be an indicator about the quality of processed data, being ineffectual and under migrated.

It is important to comprehend the Fresnel zone curvatures and the widths which in reality vary greatly contingent to several factors making it an intricate three-dimensional problem. In a two dimensional case, at a particular depth it is approximated by a simplified form as product of seismic wavelength and depth as $R \approx (\lambda \times z/2)^{1/2}$, where R, λ and z represent the Fresnel zone radius, seismic wavelength and depth respectively. The Fresnel zone width varies with depth, is small of about ten to fifteen meters at shallow depths (small wave length and distance from source) and increases to the order of hundreds of meters at depths. Since the Fresnel zone width sets the spatial resolution limit, it is important that the zone be reduced to a minimum to improve spatial resolution so as to resolve small geologic objects clearly separated from each other. This is achieved by sampling the profile with closer geophones on ground during acquisition and to a large extent in data processing by a technique called migration. Migration enhances horizontal resolution, a role similar to that of deconvolution which augments vertical resolution.

Migration

Migration technique works on mathematically continuing downward the wave field, virtually amounting to lowering of the surface geophones down up to the reflector. This reduces the distance and consequently the Fresnel zone for improved resolution. The process of migration achieves primarily (i) restoration of dipping seismic reflection events to their true geological subsurface positions and (ii) collapsing of diffractions to improve images and their continuity. Migration puts the reflected energy back where it originated from (Etris et al. 2002) and provides a reconstructed version of the true geometry of the subsurface. A schematic illustration of two dimensional migration restoring the exact disposition of a subsurface dipping segment by shifting it laterally and updip is shown in Fig. 8. It shows that the true length of the segment is smaller and the dip higher compared to its apparent location mapped in an unmigrated section. For the same reason, unmigrated or under migrated seismic sections show larger areas for anticlines and smaller for synclines. The geologic significance is that the reserves estimated for anticlines on poorly or

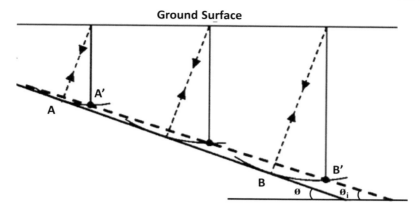

Fig. 8 Schematic illustrating seismic migration in restoring exact position of the subsurface reflector with true dip. The thick line AB represents the real subsurface reflector segment while the dashed line is its apparent position A'B' as seen in unmigrated seismic. This is because the normally reflected ray paths (arrowed dashed lines) from the real reflector are deemed arriving from vertically below the shot points and are plotted accordingly. Note the difference between the true and apparent dips and length of the reflector segment. The true dip is more and the length is shorter with the segment laterally shifted to be positioned updip to represent its proper subsurface position

unmigrated 2D data are likely to be inflated than actual. Fresnel's zone being a three dimensional phenomenon, two dimensional migrations carried out on 2D data is never perfect as it narrows the Fresnel zone in that plane only. Restoration of true disposition of subsurface features essentially require 3D migration for optimum resolution of three dimensional geologic features.

Migrated sections also preserve true reflection amplitude, creates a more accurate image of the subsurface and more importantly, enhances spatial resolution. For these reasons, migration of data is desirable even for data with flat geologic strata. Typically, the Fresnel zone widths, which are of hundreds of meters in unmigrated data, can be considerably reduced to about 10 m or so by migration. For an effective migration, however, knowledge of proper overburden velocity field and an adequate number of traces around the object, referred as aperture, is necessary for migration stack. An aperture is the spatial width over which all traces around are considered for migration, and choosing an appropriate aperture is crucial to its effectiveness. Generally, an aperture of twice the Fresnel zone width at the reflection object is adequate. However, the migration results suffer gravely near the end of seismic lines due to lack of traces recorded and the interpreter should be cautious to consider data in this part during interpretation.

For better resolution, lateral changes in reflectivity of small dimensions, migration requires finer spatial sampling on ground mentioned earlier, similar to temporal sampling used for improving vertical resolution. Take for instance the issue of imaging a small channel of 20 m width, which is often the exploration objective. Obviously, the object cannot be resolved with insufficient trace sampling of 25 m though the image with this trace spacing may be able to detect it. The channel geometry and more importantly its associated reservoir facies like channel, levee and point bar sands need to be imaged and resolved properly to characterize the reservoir and may necessitate closer trace spacing (subsurface) of no more than 10 m.

Temporal and spatial resolution may be considered somewhat similar in nature and are decided by the wavelength, which is dependent on velocity and frequency of the seismic wave. Both the resolutions depend on velocity with one exception, the temporal resolution depends on interval velocity while the spatial resolution is dependent on overburden velocity. As an example, consider a limestone bed with an interval velocity of 3200 m/s and an overburden velocity of 2400 m/s for calculating the resolution limits. Given the dominant frequency as 40 Hz, the vertical and spatial resolution limits are 20 and 15 m respectively, considering quarter wavelength as the realistic limits of resolution. It is useful for the interpreter to have some idea about resolution limits beforehand; otherwise, one may be looking for things that are beyond the capability of the recorded data to offer. It is also interesting to note that the two resolution effects are co-linked and improving one tends to better the other (Lindsey 1989). For instance, if two thin beds placed vertically or sidewise are not resolvable, a blurred image of envelop of the entwined beds would be created, whereas resolved either vertically or laterally, each of the beds can be clearly defined.

Interference of Closely Spaced Reflections; Types of Reflectors

We have seen earlier that for beds with thickness, larger than quarter seismic wavelength, reflections from their top and bottom appear as distinct and separate. However, in nature beds are commonly closely spaced in the subsurface and reflections from several beds arrive within a time spacing that is less than the length of the seismic wavelet. This leads to superposition of the reflections (Fig. 9). The ensuing interference can be either constructive or destructive and the resultant composite reflections depend on a) number and thickness of the thin beds, b) magnitude and sign (polarity) of the reflection coefficients and c) the order of positioning of the individual impedance contrasts. We may consider the behavior of three types of reflectors, namely *discrete, transitional* and *complex*, that an interpreter routinely comes across during interpretation (Fig. 10).

Discrete Reflectors

Top and bottom of thick beds with sharp impedance contrasts create distinct separate reflections with reverse signage for recording and are termed

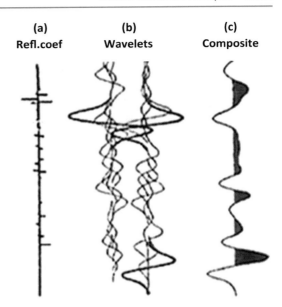

(a) Refl.coef   **(b)** Wavelets   **(c)** Composite

Fig. 9 Interference of reflections from closely spaced interfaces. (**a**) subsurface reflection coefficient series. (**b**) reflected wave let from the individual beds and (**c**) the composite reflection caused by superposition of individual reflections from the thin beds which are unresolved (modified after Vail et al. 1977)

discrete reflectors. The reflections from top and bottom appear well separated with amplitude proportional to reflection coefficients. For a zero phase wavelet, the onset of the reflection from the interface, either a peak or trough (polarity), appears at the correct time on record with respect to its subsurface depth without any time delay.

Transitional Reflectors

A transitional reflector has a gradual gradation of impedance contrasts of one signage, either positive or negative (Anstey 1977), as in a fining upward channel or coarsening upward bar sand. The interference of a succession of reflections of the same signage of impedance contrast results in a composite reflection creating a combined average wave shape. The reflection amplitude is generally weak with a low frequency appearance, the onset time is delayed with respect to the top of the formation and with unclear peak or trough to represent the contrast of beds

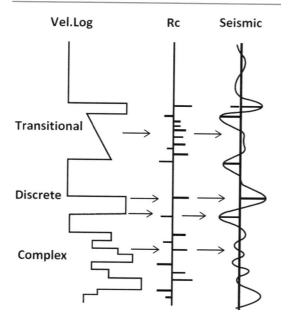

Fig. 10 Schematic showing the different types of reflectors. A 'discrete' reflector causes top and bottom reflections, resolvable with distinctive polarity and exact arrival time. The 'transitional' and 'complex' reflectors are composite events of several closely spaced beds with uncertain signage of polarity and delayed arrival time (modified after Clement 1977)

Complex Reflectors

Complex reflector is a pack of reflectors, spaced closely but with varying magnitudes and signage of impedance contrasts (polarity), which produce a complex wave form of the reflection event. The strength, onset time of the reflection and the signage of impedance contrast are difficult to gauge. Forward seismic modeling may be used as a solution to get an insight to the pattern of a complex reflection.

Innate Attributes of a Reflection Signal

A seismic trace is a log measure of disturbances (particle velocity/acoustic pressure) of waves reflected from subsurface with time. It records in

a waveform the intrinsic attributes of a reflection signal which are the amplitude, phase, frequency, and polarity, arrival time (velocity), all of which can be measured or estimated. The attributes of the reflection signal carry important geologic information encrypted in them and provide the arrival times of the geologic strata. Estimates of these rock properties from seismic waveforms and their vertical and lateral changes in time and space is the essence of seismic interpretation to predict subsurface structures and stratigraphy for petroleum exploration. The basic elements of reflected seismic wave which is the signal, are introduced here; their measurement and application are described in Chapter "Analysing Seismic Attributes". The schematic diagram Fig. 11 illustrates the attributes of a reflected wave.

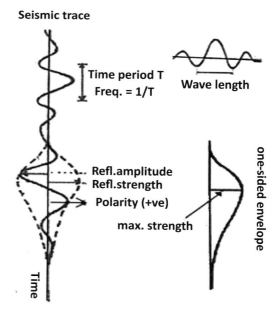

Fig. 11 The seismic signal attributes measurable from a trace, namely the time period, wavelength, reflection amplitude, reflection strength and polarity. Reflection strength is the maximum amplitude of the envelope of a composite reflection, independent of phase. Note the reflection amplitude maxima is different from the reflection strength maxima (after Anstey 1977)

Amplitude and Strength

As stated earlier, a seismic wave incident normal to an interface with an impedance contrast produces two waves normal to interface, one reflected upward and the other transmitted downward. The amplitude of the reflected wave with respect to that of the incident wave is termed the reflection coefficient (Rc) or the reflectivity. Reflectivity depends on the degree of contrast between the impedances on either side and also on the angle of incidence of the wave. For a normally incident wave, reflectivity (Rc) is expressed by the founding equation of seismic reflection method,

$$Rc = V_2\rho_2 - V_1\rho_1/V_2\rho_2 + V_1\rho_1,$$

where V_1, ρ_1 and V_2, ρ_2 are the velocities and densities of the upper and lower layers respectively. For non-normal (oblique) incidence, there will be, however, two pairs of 'P' and 'S' waves (refer Fig. 1) and the above equation for the normal reflection coefficient gets complicated and is guided by Zoeppritz's equations, discussed in Chapter "Shear Wave Seismic, AVO and Vp/Vs Analysis".

Amplitudes are measures of particle velocities or pressures and in an ideal case, for zero-phase wavelet, the maximum value at peak/trough of the wavelet pulse represents the reflection coefficient of a discrete reflector. Where the wavelet is leggy (lengthy and cyclic) and the reflection is of composite nature, as is often the case in nature, it is difficult to choose the appropriate peak/trough for calculation of amplitude to represent reflectivity. In such cases, it may be convenient to use reflection strength, which is the maximum amplitude of one side of a symmetrical envelope, centered about the reflection event. Reflection strength is more meaningful as it is independent of phase and relatively less sensitive to the factors affecting amplitude. Reflection strength may have a maximum at a phase other than at peak/trough and may indicate the nature of the composite reflection (Fig. 11) Reflection amplitude and its variations are useful tools to predict lithology of formations and their lateral changes, porosities and sometimes pore fluids as in the case of gas reservoirs. However, a crucial limitation of amplitude is its proneness to wide variance due to influence of several other factors that may not be linked to geology.

Phase

Phase may be expressed simply as the time delay with respect to the instant of start of a reflection. Phase change can be visually seen as a change in continuity of a reflection horizon manifested by a shift in the peak/trough correlated. Phase is independent of amplitude and indicates the continuity of an event which provides another useful criterion to interpret reflections. In areas of poor reflectivity, where reflection amplitudes are too weak to be manifested and correlated, phase is likely to be helpful in mapping the continuity of the reflection (reflector). Phase mapping is especially sensitive to detection of discontinuities like pinch outs, faults, fractures and angularities as well as unconformities based on 'out of phase' events. However, phase correlation needs processing of data for transforming the signal from time domain to frequency domain (Chapter "Analysing Seismic Attributes").

Frequency (Bandwidth)

A seismic wavelet, usually of one to one-and-a-half cycles duration in the beginning, changes shape progressively during propagation and becomes long and cyclic (leggy) with passage of time. The pulse width of a wavelet on the seismic record in time (time period) provides an estimate of its dominant lowest frequency, and it becomes wider with depth during propagation indicating lowering of frequencies caused by attenuation. The bandwidth is a measure of the width of a range of frequencies in the wavelet, measured in hertz and is the key to quality of reflection. Bandwidth of a wavelet decides the duration time (width) of the changing wavelet corresponding to depth intervals, reliant on the velocity and defines the vertical and co-linked lateral seismic resolution. A broad bandwidth consisting of both low and high frequencies is thus essential to provide quality seismic images. Ironically, the frequency ranges behave in contrasting manner. The lower frequencies in the spectrum help in deeper penetration of energy but have poor resolution power whereas the higher frequencies have poor depth penetration but provide higher resolution to delineate thin beds. Unfortunately, during propagation of the wave, the earth attenuates the high frequencies and hampers desired resolution at depths.

Because frequency is affected by propagation phenomena like absorption and transmission in the subsurface, frequency variance can provide valuable geologic information. Generally reflections dominant with high frequency looks indicate thin layers of strata at shallower depths whereas relatively low- frequency dominated reflections indicate older and harder rocks (i.e. Pre-Tertiary) at deeper depths. The differences in frequencies of groups of reflections evinced on data can sometimes be strikingly clear to suggest unconformities. Experienced seismic interpreters are familiar with such clearly discernible decrease in frequency of reflections from the top to bottom of a typical seismic section. Bandwidth, amplitude and phase create the shape and form of a signal, and the individual components can only be measured and analyzed by detailed spectral analysis, discussed in Chapter "Analysing Seismic Attributes".

Polarity

Polarity is an attribute which represents signage of reflectivity and is different from phase which is an intrinsic property of a wave. Polarity of a reflection signal is crucial in seismic-well ties for correlating reflection horizons. It helps identify lithologies and is the mainstay for analysis of high amplitude seismic anomalies (DHI) for detection and validation of hydrocarbon in reservoirs and fluid contacts. Hydrocarbon bearing DHI anomalies, 'bright', 'dim' and 'flat spots' are essentially characterized by the signage of reflection polarity (Chapter "Direct

Hydrocarbon Indicators (DHI)") and it is important the seismic analyst is mindful of the polarity displayed in the seismic data before interpreting DHI anomalies. Polarity expresses reflectivity of a bed interface and is considered positive if the impedance of the rock below is positive (a hard rock underlying a soft rock) and negative, the other way round.

Polarity Display Conventions, Acquisition Source Wavelet

In processed seismic data, polarity can be displayed in different ways depending on the conventions followed by individual companies or/and the interpreter's preferred choice. There are basically two standard display conventions mostly followed, the SEG in USA and other countries and the Europa, in European countries. The display conventions are essentially based on the type of source wavelet used in seismic data acquisition. In processed data, the SEG normal polarity display for minimum phase source wavelets, generated by dynamite and air-gun sources, compression (+ve Rc) is represented by trough (white/red) and rarefaction (−ve Rc) by peak (black/blue). However, the SEG normal polarity convention for zero-phase wavelet generated by Vibroseis source is opposite; the compression (+ve Rc) is represented by peak (black/blue) and the rarefaction (−ve Rc) by trough (white/red). The European normal polarity convention is just the opposite of SEG normal polarity for both type of source wavelets, i.e., for minimum-phase, compression is denoted by peak (black/blue) and rarefaction by trough (white/red) and conversely, compression by trough and rarefaction by peak for zero-phase source wavelet. The SEG and Europa polarity display conventions are shown in (Fig. 12). However, display option of reverse polarity exits in both SEG and Europa for interpreters, which is just the opposite of normal polarity. Thus, there can be eight display options for polarity, namely the SEG and the European normal and their reverse, for each of the two types of source wavelet, the minimum and zero phase. This

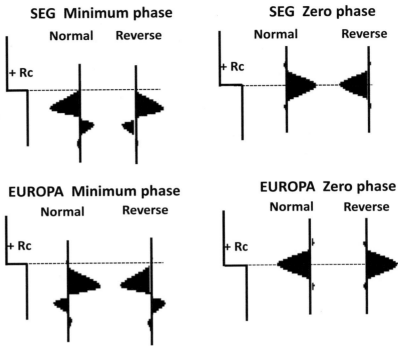

could be indeed confusing unless the polarity display in seismic sections are clearly indicated to link the signage of the reflection, for example, stating that positive Rc (+ve) represents peak/trough along with the color code (black/red).

But how and why is the polarity display convention tied to the type of source wavelet used in seismic data acquisition? The reasons for opposite polarity display for the minimum and zero phase wavelet for SEG normal polarity is explained and illustrated (Fig. 13). The inherent presumption is that the minimum phase source wavelet is a '*causal*' wavelet which causes the wave motion to begin after the onset of the wavelet (Fig. 13a). This causes the reflection to appear at the start of the first displacement and with a time lag. However, it is different for zero-phase wavelet, generated by Vibroseis source in acquisition. The zero phase wavelet known as the *Klauder wavelet* is different as it is embedded in the source wavelet and no longer represents the observed physical quantities such as the geophone amplitude as in explosive or air-gun

sources. The zero phase *Klauder wavelet* is considered '*acausal*' (non-causal) wavelet which means the wave motion begins before the onset of the wave and the maximum displacement located at zero-lag (Fig. 13b). Universally, compression (positive reflectivity) is recorded as a negative number on tape during acquisition for minimum phase source wavelet and is retained as a trough in processed data without change (Fig. 13c). But for zero phase wavelet the compression is recorded as a positive number on tape during acquisition and is retained as a peak (Fig. 13d) which is opposite to polarity displayed for minimum phase data.

Accurate picking of polarity is important for locating the disposition and nature of the strata in the subsurface. Picking of reflection polarity is simple and straight forward in case of discrete reflectors, but is difficult in transitional and complex reflectors where superposition of reflectors create composite reflection events leading to obfuscation of individual reflection events and loss of polarity information. Noise in data also acts as a deterrent for clear

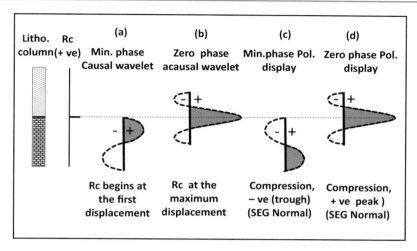

Fig. 13 Illustrating minimum (*causal*) and zero phase (*acausal*) wavelets and their link to polarity display convention in SEG (Normal). Note (**a**) the onset of reflection maxima delayed as reflection starts at the first displacement for minimum phase 'causal' wavelet and (**b**) the amplitude maxima without delay for zero phase 'acausal' wavelet. (**c**) compression conventionally recorded on tape as a negative number (trough) and is displayed as trough in processed data whereas (**d**) for zero phase wavelet, it is recorded positive and displayed as peak (courtesy: Satinder Chopra, Calgary)

identification of polarity. Processing techniques such as deconvolution and zero phase processing can help to some extent in estimating appropriate polarity of composite reflection events.

Arrival Time (Onset of Reflection)

It is therefore desirable the seismic interpreter is aware of the source used in data acquisition and the polarity convention displayed by the processing centre, prior to interpretation of data. A simple but important consequence of polarity information is the issue of picking the correct arrival times of reflected events which in many cases are not discrete but are of complex nature. The concern often is about which phase of a reflection is to be picked on seismic time sections, the peak, the inflection point (zero crossing) or the trough. This is crucial as the phase is important for well calibration and correlation while its arrival time is vital for depth estimation in interpretation. For instance, for minimum phase source wavelet used in offshore (air-guns) and on land (dynamites) in most cases, the reflection beginning at the leading edge displays the maximum amplitude of peak/trough, which happens to be delayed in time corresponding to the true depth of the object. Despite identifying the proper polarity, the picked time would therefore warrant appropriate time corrections for accurate depth prediction. For zero-phase wavelets, however, since the displacement maxima corresponds exactly to the depth of interface without time delay no such problem exists. Interpreters therefore prefer to work on zero-phase data which can be achieved in processing by phase shifts to minimum phase recorded data.

However, in offshore data a simple observation can provide clue to the polarity information. The sea bottom is generally a strong reflector with positive reflectivity (+Rc), being the interface between water and sediments. The energy source is known to be air-gun which generates a minimum phase wavelet. The polarity display of the sea bottom reflection therefore provides the clue to polarity display - if it is a trough, it is normal SEG polarity (Fig. 14). A peak would indicate otherwise, that the data is either zero-phased or reversed in polarity at the processing centre. However, there is also a catch, often the first strong signal recorded from the sea bed

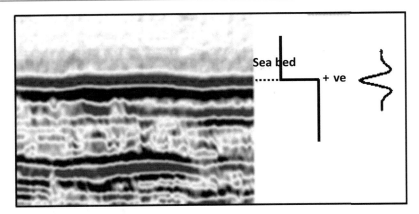

Fig. 14 Illustrating identification of polarity convention from seabed reflection in offshore data The sea bed reflector between water and sediments has positive Rc and shows strong reflections. The air-gun source is minimum phase and by SEG Normal convention it would show negative amplitude (trough, red). If the polarity is seen as peak, the data is processed zero-phased or with SEG reverse polarity (image courtesy, ONGC, India)

reflection are muted during processing which makes it difficult to determine exactly the polarity of the sea bottom reflection.

Despite correct reflection phase and time picked and velocity used, there can be mismatch between the times converted depth and the actual subsurface depth. This is because the picked reflection time may have been delayed due to acquisition and processing systems that behave as filters and introduce time lags. The induced delays may ultimately range from few to several milliseconds, depending on type of seismic data (2D/3D). Seismic analysts often find such time shifts in tying a particular reflection phase in different vintages of seismic, especially in 2D data, due to varying recording and processing parameters. This may be reconciled by advancing the reflection time by a negative correction though the exact amount would be a best guess process.

Velocity

Velocity is an important seismic property, not only to estimate depths of formations, but also to provide vital information on subsurface rock and fluid properties. Basically, velocities are of two kinds, the *overburden* or vertical average velocity, and the *interval* or formation velocity. The two velocities are interrelated, knowing one can lead to compute the other. Other types of velocities, stacking (NMO), root mean square (RMS), migration and instantaneous velocities are also briefly described.

Average Velocity

Average velocity is the true vertical velocity used for conversion of reflection times to depth, and is the most important element in the exploration gamut. The true vertical average velocity, besides used for determining depth to geologic objectives, helps deduce the crucial interval velocities accurately for stratal layers and more importantly for proper prestack depth migration of data.

Stacking Velocity

Stacking velocity, also known as normal move out velocity (NMO) is an overburden velocity computed mathematically during velocity analysis process for normal move out correction of the multi-offset seismic traces used in common depth point (CDP) technique. The NMO corrections are for adjusting the geometrical effect of the varying offsets so that the traces are transformed to normal incident time for stack with maximum amplitudes. Stacking velocity is an apparent velocity recorded along the spread of ground geophones and is affected by factors such as dips of strata and acquisition spread lengths. Stacking velocities are usually higher (by about 6–10%) than true vertical velocity.

RMS Velocity

RMS velocity is root mean square (RMS) velocity and is another way to denote stack velocity. Assuming that the subsurface layers are parallel and horizontal, RMS velocity permits to deduce mathematically the layer interval velocities by a formula known as Dix's formula. In the absence of well velocities the RMS velocity, after appropriate correction, is used to predict top, bottom and thickness of geologic formations and also for migration in data processing. The lithology and other rock properties can be also inferred qualitatively from interval velocities (formation velocity) calculated from RMS (stack) velocities. Unfortunately, the above assumptions about the beds being flat and parallel are never met in nature and consequently the derived interval velocities have the inherent inaccuracies. Grossly speaking the NMO, stack and the RMS velocities genetically belong to a group of velocities almost similar to one another.

Migration Velocity

Migration velocity is the velocity used for time migration of seismic data to accurately locate the true subsurface reflecting points below the shot point. Similar to stacking velocity, migration velocity is an overburden velocity that provides the optimum imaging to produce relatively clean and accurate seismic images that helps predict structural configuration of geologic formations, their depth and rock properties. Migration velocity takes into consideration both the horizontal and vertical components of overburden velocity in contrast to vertical velocity used for depth conversion. Commonly, the migration velocities are stack (RMS) velocities used with some modifications. It is usually lower than stack velocity but tends to equal the average overburden velocity when data is optimally migrated.

Interval Velocity (Vint)

Interval velocity is the velocity of a formation (layer) and is called formation velocity. Interval velocities of a number of formations can be integrated to compute vertical overburden velocity and the other way round, from a given a series of layers with interval velocity the vertical velocity can be calculated. Interval velocities are generally calculated from RMS velocity and are important to predict lithology and rock-fluid properties of the layer, such as porosity and fluid contents. However, interval velocities are highly sensitive to interval of thickness for which they are computed and also to accuracies in velocity picks during seismic velocity analysis in data processing that provides the stack or RMS velocity.

Instantaneous Velocity (V$_{inst}$)

Instantaneous velocity is a velocity at which a seismic wave propagates at a point within the interval of a formation, similar to interval velocity. However, instantaneous velocity is slightly different from interval velocity. Instantaneous velocity within a given interval changes with depth defined by a gradient, whereas the interval velocity remains constant without change in the interval. While interval velocity denotes the average formation velocity, the instantaneous signifies the finer details such as the layer interval velocities within the formation. The instantaneous velocity concept can be best realized as being closest to continuous velocity log (CVL) computed from sonic log by converting the slowness (μs/ft) to velocity (m/s). The change in instantaneous interval velocity with depth, the gradient and the constant interval velocity is illustrated in Fig. 15. Important applications of instantaneous velocity include computing synthetic seismograms wherein the product of sonic log derived instantaneous velocity and the density provides the impedance log as the input. More importantly, the Instantaneous velocities are used in building velocity models for depth conversion and migration and are discussed in Chapter "Seismic Interpretation Methods".

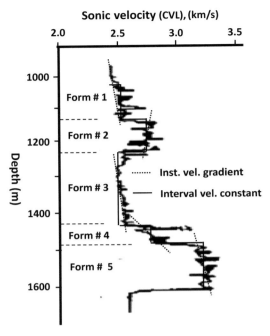

Fig. 15 Shows sonic instantaneous velocity curve (CVL) illustrating instantaneous and Interval velocity. Interval velocity is constant for formation while instantaneous velocity varies within It with a gradient. Interval velocity shows the gross formation velocity while instantaneous denotes the layer velocity with in the formation (courtesy ONGC, India)

Seismic Display

Display Modes

The visualization of seismic data is an integral part of interpretation and as such, it is important that the processed seismic data be displayed in suitable graphic modes and scales. Nonetheless, it depends to a large extent on the objectivity of the interpretation and the perception and creativity of an individual interpreter. Generally, data are displayed in any one of these modes, wiggle trace, variable area, and variable density or in combination.

- Wiggle trace is a log of reflection amplitudes with time and makes it handy to interpret geologic information from the variability in the waveform shape (Fig. 16a).

- Variable area (VA) and wiggle displays are wiggles shaded with bias, however, make reflection events appear more consistent and convenient for correlation by reflection character, the wiggle shapes providing better geologic information (Fig. 16b).

- Variable density (VD) shows reflection strength, and displayed with color, provides better relative standout and continuity of reflections. VD sections, though more commonly used showing better horizon continuity, do not show the waveform shapes which embed significant geologic information (Fig. 16c).

- Combinations of variable area with wiggles and at times may be a preferred display for interpreting stratigraphic details (Fig. 16d).

Though the work stations provide different modes of display, interpreters generally use variable density sections due to better apparent continuity of reflections and their amplitude stand-outs that are conveniently used for extraction of seismic attributes for display (Fig. 16c). But the continuity in correlation can at times be misleading in many geologic settings. For instance, in continental to fluvio-deltaic depositional environments, where fast and frequent facies variations are likely to occur, one would normally expect discontinuous and patchy reflections and not wide-spread continuity of horizons. Seismic reflection horizons with perceived good continuity may in such cases, do not properly represent the subsurface geology and seismic interpretation without the geologic set-up can be highly flawed. Use of seismic display in wiggle with variable area mode may be preferred for correlation of seismic reflections. Horizon correlations guided by wave form characters permit better sensitivity to perceive verticolateral variabilities in the wave shapes which carry the subsurface geologic information (Fig. 17). Wiggle with variable area displays are also desirable for calibrating synthetic seismograms/corridor stacks with seismic which can clearly show the quality of matching of events.

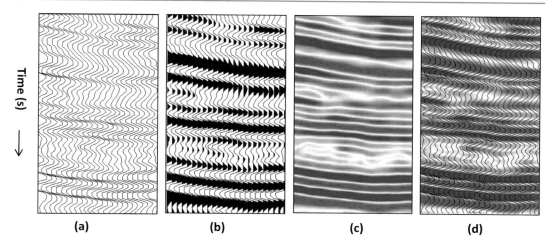

Fig. 16 Seismic segments showing types of display modes. (**a**) wiggle, (**b**) wiggle and variable area, (**c**) variable density, (**d**) combination of wiggle and variable density. Note the lateral changes in wave form seen clearly in wiggle and variable area mode (**b**) but not so clear in (**c**). Lateral variability in wave forms provide valuable geologic information

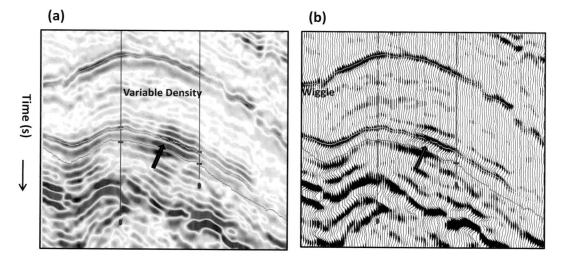

Fig. 17 Seismic image in display modes showing comparison of (**a**) variable density and (**b**) Wiggle with variable area modes. Reflection standouts and continuity are seen better in the variable density, but does not show the changes in wave form and misses the important geologic information they carry. Wiggle and VA mode shows clearly the variations (the trough marked by arrow) and help proper correlation based on reflection character and also offer geologic information from the lateral change of wave forms (image: courtesy, Hardy Energy, India)

Color display is known to increase optical resolution leading to better visual discrimination of features and is widely used. The selection of suitable color and its encoding depends on the artistic attitude and geologic perception of the interpreter so that the seismic vertical sections provide a look close to natural subsurface geology. Assigning colors in a spectral progression is usually preferred as it enhances the relative magnitudes better for visualization.

Plotting Scales (Vertical and Horizontal)

Plotting scales are extremely important in data display, as reducing or stretching the scales impacts visualization of geologic objectives. Visualization of images is important as it is an integral part of mind and the image interpretation depends often on what the mind dictates. For the same reason, the plotting scales are also to be suitably chosen depending on the objectivity of the interpretation. Horizontally compressed sections promote perceiving better continuity of events and the subtle dips appearing stronger. Stretched sections, on the other hand, appear to deteriorate reflection continuity with flattening of the dips. Vertically stretched and horizontally compressed sections improve the dip effect and make minute dips noticeable better. Low angle faults with small displacement, lowly dipping progradations, gentle pinch outs and terminations etc., which are subtle but important as exploratory objects, can be made to look more conspicuous by suitably adjusting the vertical(time) and horizontal

display scales to be picked as anomalies (Fig. 18). Horizontally compressed (squashed sections) and vertically reduced (reduced time scale) sections play a very useful role in studying regional basin scale geology, helpful in understanding evolution of basins for evaluating hydrocarbon prospectivity. Choice of scaling depends entirely on the interpreter depending on the length of profile and the depth (time) recorded. Usually squashed sections are plotted every fourth trace with proportionate reduction in time scale so that long segments of profiles can be conveniently viewed in one frame. This offers the advantage of viewing the entire geology, from the deep basement to the surface for assessing the tectono-stratigraphic evolution of the basin. However, vertically stretched sections are at times used to magnify details of important target windows for mapping the objective. Each geologic object thus needs appropriate scales of display for the target to standout clearly and may need experimenting for choosing the best judicious combination of both the vertical (time) and horizontal (trace) scales along with the mode of display.

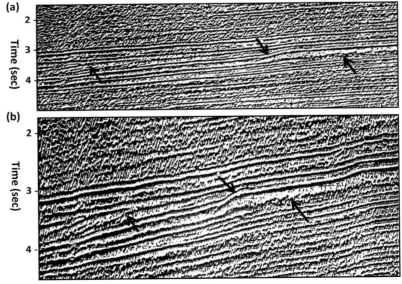

Fig. 18 Seismic segment illustrating visual effects of plotting scales.
(**a**) horizontally stretched and vertically compressed and
(**b**) horizontally compressed and vertically stretched. Note the subtle dipping features clearly seen in compressed section (**b**) not well discernible in expanded scale (**a**). Arrows point out the anomalies (courtesy, ONGC, India)

References

Anstey AN (1977) Seismic interpretation- the physical aspects, records of short course "the new seismic interpreter. IHRDC 2-109 to 2-111A, 3–1 to 3–19 & 3–65 to 3–85

Clement WA (1977) A case history of geoseismic modeling of basal morrow-Springer sandstone. AAPG Memoir 26:451–476

Etris LE, Crabtree NJ, Dewar J (2002) True depth conversion: more than a pretty picture. CSEG Recorder 26:1–19

Lindsey JP (1989) The Fresnel zone and its interpretive significance. Lead Edge 8:33–39

Meckel LD Jr, Nath AK (1977) Geologic consideration for stratigraphic modeling and interpretation. AAPG Memoir 26:417–438

Sheriff RE (1977) Limitations of resolution of seismic reflections and geologic detail derivable from them. AAPG Memoir 26:3–13

Vail PR, Todd RG, Sangree JB (1977) Chronostratigraphic significance of seismic reflections. AAPG Memoir 26:99–116

Widess MB (1973) How thin is thin bed? Geophysics 38:1176–1180

Seismic Interpretation Methods

Abstract

Seismic interpretation conveys the geologic meaning of data by extracting subsurface information from it. Type of interpretation depends on the geologic objectives linked to the phase of exploration and on the kind of available data, 2D/3D, its grid density and quality. 2D seismic data interpretation can be typed as structural, stratigraphic and seismic stratigraphy. Workflows for the 2D interpretation types are described in this chapter, underscoring the basic principles along with some of the application shortcomings that crop up during normal interpretation practices. Techniques of seismic calibration, horizon correlation, preparing seismic maps, and velocity modelling and depth conversion are elaborated with examples and illustrations. Seismic stratigraphy is typically a regional geologic interpretation of seismic 2D data, and is described with schematics and real seismic image illustrations. The difference between seismic stratigraphy and stratigraphic interpretation is highlighted.

It is necessary the seismic interpreter stresses on the consequences of interpretation results that impact the geological and engineering issues in the exploration project. This is a process which adds value to interpretation and may be termed as data evaluation. Data evaluation looks ahead beyond the usual outputs of routine interpretation and assists in review and assessment of economic viability of prospects which enables management to strategize exploration policies. This is emphasized by citing examples and illustrations in the chapter.

Petroleum exploration is a high-cost and high-risk intensive venture, which demands important subsurface geologic information, as precise as possible, with minimal prediction upsets. Interpretation of seismic data offers these decisive geologic inputs for exploration undertakings. Providing reliable seismic predictions requires a synergistic and systematic approach to analysis of seismic and all other related data by experienced and skilled persons. It is desirable to have all relevant and accessible data sets organized in a multi-disciplinary database prior to the start of a comprehensive seismic interpretation work flow. Multiplicity of data sets and multi-disciplinary data types, though are a challenge to handle, improves the scope for better synthesis and dependable evaluation. It is of course essential that the seismic interpreter, has a good understanding of petroleum geology, petrophysics, and reservoir engineering in addition to in-depth knowledge of seismic technology which includes seismic data acquisition, processing and interpretation(API) and related other geophysical techniques.

Interpretation reveals the geologic significance of seismic data by extracting subsurface information from it. Seismic interpretation, in the context of petroleum exploration, however, should not be limited to only offering the geophysical results but needs to be geologically inclusive to address the exploration problem at hand. Interpretation thus needs to be logically stretched to include the geologic and engineering implications of the seismic inferences on the exploration venture so that it eventually helps management in taking sound techno-economic decisions. This is termed data evaluation, a process that adds value to interpretation and must look ahead, beyond routine interpretation, into the success of the entire exploration project at hand. Emphasis therefore may be put on data evaluation, which assists strategizing exploration policies. For example, detecting and mapping faults in a prospect is not an end in itself of interpretation; it is more important to evaluate their significance in impacting the potential of the prospect in terms of the roles the faults may play in accumulation and production of hydrocarbons.

2D Seismic interpretations can be of different types depending on the geologic objectives linked to the particular phase of exploration and the available seismic data type, grid density and quality. The petroleum exploration and production (E&P) activity cycle generally starts in its first phase with analysis of seismic and other geophysical data that leads to generate a prospect for drilling well(s) known as exploratory well. In case of a discovery, more exploratory inputs follow in the onset of the second phase of exploration. This may include acquiring more and higher resolution close-grid seismic data to better delineate and characterize the reservoir needed for identifying suitable locations for drilling, the delineation/appraisal wells. Depending on economic viability (Chapter "Seismic Stratigraphy and Seismo-tectonics in Petroleum Exploration"), appropriate development plans are initiated leading to the third and final phase of field development by drilling production wells to start production. This has been traditionally a normal practice, but explorations may be carried out differently in exploration endeavors contingent to several extraneous factors ranging from geopolitical to commercial

causes. However, in classical sense, each phase, the prospect generation, delineation and development, may be linked, albeit historically and in an orthodox way, to the kind of interpretation contingent to type of data and the objective at hand in the initial exploration phase. Higher the order of exploration phase, more the need to have better resolution data followed by more intensive level of interpretation. Having said all that, it may also be added that seismic evaluation does not necessarily end with the development phase but carries on till the field is depleted and abandoned.

This chapter deals with 2D seismic data interpretation which may be categorized as below.

- Category I: Structural Interpretation (2D data) – first level regional interpretation, mainly structural in nature, and usually done in the first stage of exploration.
- Category II: Stratigraphic Interpretation (2D/3D/4D data) – higher-level synergistic interpretation that provides stratigraphic information including rock and fluid properties during exploration/delineation and production stage.
- Category III: Seismic Stratigraphy Interpretation (2D/3D data) – regional geologic interpretation of depositional systems and tectonic styles for basin evolution and evaluation, and is carried out mostly in initial stage of exploration.
- Category IV: Seismic Sequence Stratigraphy Interpretation (2D/3D data) – a more detailed version of seismic stratigraphy and an integrated stratigraphic interpretation of log and core data with or without seismic and can help in exploration, delineation, and development stage.

With relatively more close-grid and better quality 2D data, a comprehensive interpretation, providing details about subsurface structures and stratigraphy may be attempted in the beginning in the early exploration phase. Sophisticated softwares, presently available are widely used to facilitate seismic interpretation using advanced workstations and large integrated databases. Nonetheless, it is essential that the principles and methods of interpretation and linked framework be properly understood by the interpreter behind the machine so as to extract maximum meaningful geologic information through a man-machine interaction. Interpretation and evaluation of high density –high resolution 3D seismic data is discussed in the Chapter "Evaluation of High-resolution 3D and 4D Seismic Data".

Category I: Structural Interpretation (2D)

Structural interpretation of 2D seismic data is generally practiced in virgin or less explored areas. It primarily involves mapping structural features and their preliminary evaluation through mapping of subsurface stratal boundaries, the reflectors, called seismic horizons. The initial data to work with generally consists of a coarse grid ($\sim 4 \times 8$ km spacing between survey lines), routinely processed seismic data with no or sparse well control. The interpretation, in the absence of geologic data from either nearby wells or outcrops, may be more geophysical in nature. Nonetheless, the interpretation delivers useful information about the depth of basement and its configuration, the paleo-highs and lows and the faults, the order of total sedimentary thickness in the basin and the structural features present within. More importantly, the interpretation identifies the individual sedimentary units (formations) in terms of depth, thickness, lithology, and their structural forms and types including the faults. These are the initial important inputs to reconnoitery exploration in virgin areas for generating hydrocarbon prospects. Interpretation also leads to reveal and prioritize the potential areas for further exploratory inputs such as close-grid 2D or 3D data acquisition. The major steps involved in the work flow are: (1) identifying horizons (reflectors) for correlation; (2) picking and posting time values on a location (base) map; (3) contouring the horizons with faults and

(4) creating time structure and interval maps for the horizons for showing geologic details.

Identification of Horizons and Correlation

The deepest event (horizon) seen on a seismic section is usually considered the basement reflection. This reflection is generally characterized by low amplitude and low frequency, and is often discontinuous and punctuated by a number of faults. This horizon sometimes is termed a *technical* or *acoustic* basement by the interpreter where the fundamental Precambrian basement (Archaeozoic) is believed to be deeper but not seen in seismic. Reflections, or horizons, that are continuous and present over a wide area and can be easily correlated by their strong character are known as seismic "markers", analogous to a geologic marker bed.

Identification of Horizons

Identification or picking of reflections for correlation is usually based on some criteria. Picking of horizons includes the basement top, the markers (if present) and the other horizons that show good lateral continuity and discordant dip attitudes with respect to each other. Discordance in the horizon attitudes signifies unconformities and are important for correlation and mapping. Seismic horizons that are parallel are generally not picked as they do not add to structural information, mapped from one of the horizons. The nonconformable horizons are preferably picked first on a dip-line as it is more convenient for correlations because of likely clearer lateral continuity and dips of reflection events and then extended to all other lines in the area.

Correlation of Horizons

Correlation of a horizon involves linking the reflection from one seismic line to another based on reflection character thereby making sure that the stratal surface followed is the same over the area. Reflection character considered for correlation comprises of amplitude, phase (peak/trough), frequency, waveform and the dip attitude of the horizons (Fig. 1). Sometimes extending the continuity of a reflection may be uncertain because of change in the character due to noise, lithological variations and faults amongst other reasons. This is generally dealt, as per need, by resorting to extension of continuity by jumping across the poor and no reflection segments, known as '*jump correlation*'. The continuity of a reflector may also be forced intuitively across the uncorrelatable portion, a process known as '*phantoming*'. The decision to continue correlations beyond the reliable limit using *jump correlation* or forcing *phantoming* across poor/no reflection zones, data gaps and faults, depends on the interpreter's judgement and compulsions of exploration need at hand.

Phantoming and jump correlations of horizons can be highly subjective with reliability depending largely on the individual skill and experience of the interpreter. Despite being highly subjective and tentative, these type of correlations can at times be useful where mapping of structural attitude of the horizon is the prime aim of interpretation. Since the correlations can be highly arbitrary, influenced by the interpreter's bias and judgment, it is desirable to use some kind of a code to denote the confidence level. Qualifying the correlations as fair, poor, questionable, etc., by designated symbols (such as '- -', '??' etc.,) is important to indicate the order of uncertainty so that appropriate decisions can be taken accordingly by the management.

During correlation, distinction must also be made between steep dips and faults in the horizon, as dips and faults have different geologic significance. Though there may not be difference in the spot time picks of the correlated horizon and its structural configuration in a map, a fault indication may have other consequences linked to hydrocarbon potential of a prospect (Chapter "Seismic Stratigraphy and Seismo-tectonics in Petroleum Exploration"). Faults, therefore, need to be identified, picked carefully during correlation for proper mapping. The legitimacy of a horizon correlation being from the same geologic strata over the entire area is generally established by matching the reflection times and more importantly the character at the crossing point of two or more lines. This is accomplished through a process known as '*loop tie*', similar to that used in cartographic surveys, which ensures same time values at the start and end of a loop made through different lines. However, this is not

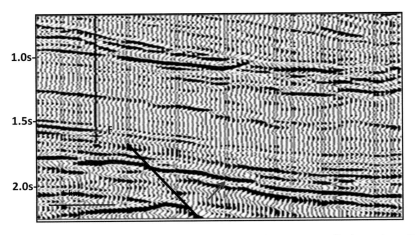

Fig. 1 Seismic segment illustrating correlation of horizons based on reflection character. Reflection character includes phase (peak/trough), amplitudes, frequency and dip attitude. Note the reflection character of 'trough (arrow) with peaks on either side' is continuing up-dip till the fault, beyond which it loses the character though the top peak with similar amplitude continues further. Ideally the fault limits the continuity of the horizon, however, further correlation, up-dip can be done by 'phantoming', which could be subjective. (*Image courtesy of ONGC, India*)

always a foolproof method as despite the matches at cross points, there can be miscorrelation in between the wide-grid data that can impact accuracy of mapping.

'Mis-ties' at crossing points in 2D data are fairly common and are mainly due to deficiency of two dimensional migration that fails to restore the subsurface reflection points to its true subsurface location accurately. Another reason, however, can be due to navigational inaccuracies in ground locations. Mis-ties may be considerable in seismic data belonging to different acquired and processed vintages, using dissimilar recording and processing parameters including cartographic and navigational errors. It is important that the mis-ties at the cross points in the loops are suitably adjusted to the least-time of the horizon that is seen in the vintage sets to prepare maps. Computer softwares, however, automatically take care of misties in the work station before attempting contouring but may not handle reasonably the large discrepancies. In cases of considerable misties between two vintages, a good choice for the interpreter may be to use only the vintage set which shows better correlatibility for mapping purpose. However, other sets of poor data may not be discarded outright as valuable information regarding the dip attitudes of the horizons from this data can still be useful to incorporate for improving structural details better.

Correlation may be considered elemental, but it is critical initial step for preparation of the all-important seismic map, which ultimately guides the management in taking major exploratory decisions including drilling of wells and estimating reserves. Lapses in proper correlation of horizons, particularly across faults can affect accuracy of the map. Unknown to the exploration manager, the map may be accepted as accurate, based on which flawed decision may be taken. As stated earlier, in this context, qualifying the level of confidence in picking and correlating the horizons including the faults by using coded terms cannot be overemphasized.

Recent techniques allow correlations of horizons and picking times by computer in 'auto track' mode which is efficient, fast and accurate.

But accuracy does not necessarily assure correctness and 'auto tracking' has its shortcomings due to the following reasons:

- Lateral changes of polarity in events due to change in facies and rocks properties that change the reflection character may not be recognized automatically.
- Thin interbedded sandstones and shales may not be accurately picked in auto mode.
- In geologically complex and seismically poor areas with noisy, poor data quality and frequency splitting of reflection, it may cause problems for the auto mode to pick the correct phase of the horizon.

However, options are available to track the horizons manually and in such circumstances, it may be necessary for an interpreter to intervene and interact with the machine for more reliable correlations. Leaving the task to be performed by the machine unsupervised may cause lapses leading to flawed maps. Auto tracking has options to choose a reflection phase such as the peak, trough or the zero crossing and it is important the seismic interpreter identifies (i) the auto tracking phase that may best assist correlation of the desired object horizon and (ii) controls the limits of auto tracking approach in a given seismic dataset.

Correlation of seismic horizons in workstations also automatically pick the amplitude values, which can be mapped along with the horizon. Since the amplitude depends on properties of the rocks on either side of the interface, lateral variability in amplitudes can provide valuable information on stratigraphic details pertaining to subsurface geologic formations. Change in reflection horizon amplitude signifies change in facies and its continuity, defining the limit of extent of the strata. Lithologies of layers can be easily and conveniently predicted by analyzing the interval velocities (formation velocity) between two horizons, derived from seismic stack (RMS) velocities. Combined studies of interval velocity and horizon amplitudes can also indicate qualitatively formation porosity.

Contouring and Maps

The time values picked are posted with faults on the base map and are contoured to produce time structure maps of horizons. Before contouring, two simple but important parameters, the scale and contour interval, need be appropriately chosen. The map scale is selected based on size of data grids; for example, small scales are preferable for widely spaced data and larger scales for closer grids. Intervals at which contours are to be drawn may have to be decided depending on data reliability linked to degree of scatter observed in the timing of picks at crossing points. A normal practice for choosing contour interval is at least about two times the scatter error. For example, a contour interval of 20 ms may be a good choice for drawing contours if the overall cross-tie errors at the point of intersections are below 10 ms. Smaller contour intervals reveal more structural details but must be commensurably supported by quality and density of data (grid size) which is generally not met in 2D data. Maps prepared with contour intervals smaller than that supportable by data ostensibly to exhibit finer details, may indeed be misleading and may be dealt with caution.

Contouring time values is not just a trivia of mathematically interpolating grid data and mechanically joining the equal values but is much more than that. Seismic structure contours represent a particular chronostratigraphic horizon, geologically a stratal surface and are different from surface topographic contours which portray a relief map. Therefore, seismic structure contours are expected to depict structural trends and depositional geometry of geologic bodies and require supervision by skilled and experienced persons. Special efforts may be essential where there are large number of faults present which require appropriate delineation and mapping. Particularly for antithetic (fault planes showing dips opposite to bed dip) and reverse faults, it needs careful adjustment of contours across the faults to show their correct displacements, a factor that can have serious geologic implications in evaluating hydrocarbon potentials of a prospect. Faults picked on seismic lines, but delineated erroneously in maps, especially in geologically difficult cases and widely spaced 2D data, may create pseudo fault-closure trap prospects, leading to unsuccessful drilling.

Contouring by hand is an arduous, time consuming process but importantly represents interpreter's geologic thinking. The reflection time values at grids may not be all that precise due to several snags; often a few values not fitting into the shape of things, requires going back to data for a careful check. The time values sometimes are modified and even discarded by the interpreter in contouring to be geologically meaningful. Modern interpretation softwares automatically import the correlated time picks to the base map for contouring. Computers prepare contour maps quickly and efficiently, but cannot perform the task of rechecking/revising the grid values, an option available in manual mode. Computer generated contours at times may be too mathematical to be geologically realistic as the computer algorithms do not have the kind of intelligence and insight, the interpreter has, to analyze and interpret data. For 2D seismic data comprising of relatively large and irregular grid data and in complicated geologic set-up with faults of different types, machine contouring may at times be found wanting. An ideal situation may be a hybrid approach, to get the preliminary contouring done by the machine to save much needed time and then to carry out manual editing to get best results for depicting geologic structural trends and features. More of this is discussed and exemplified later under 'Stratigraphic interpretation'.

Seismic maps are the ultimate end-products of 2D interpretation and are of different kinds, outlined below.

Time Structure Maps

Time structure maps are horizons contoured in time and are the prime outputs of structural interpretation, which define the geometries and trends of subsurface geologic horizons. It is a good practice to cross check the structural features and trends mapped in plain view with the vertical seismic sections for confidence in interpreting the geology.

Depth Maps or Structure Maps

Depth maps are prepared by converting the time values of the horizons to depth. The vertical average velocities for depth conversion are computed either from nearby well velocities or from seismic stacking velocities, after applying suitable corrections. This is dealt in more detail under 'stratigraphic interpretation'.

Relief Maps

In the event of uncertainty in the correlation of reflection belonging to the same stratal surface, and with phantoming used, it may be appropriate to name the map as a *relief* map to represent as top of the envelope of the time transgressive surfaces. An example can be the annotation 'structure map on top of basement'; often based on unclear and dubious correlation of reflection from the top of basement which is an unconformity surface. A more appropriate annotation may be the term 'Basement Relief' map.

Isochrones/Isopachs

Maps showing formation intervals in time and in depths between two horizons in time and in depths are termed isochrones and isopachs respectively. The interval maps provide formation thicknesses and their thickening and thinning provides important geologic clues to depositional and tectonic setup for the older formation in the basin. Thickness maps, however, are more sensitive to data errors because of chances of inaccuracies involved in mapping two horizons, and are generally prepared with larger contour intervals.

Interval maps may also preferably include faults affecting the horizons with suitable annotation by standard symbols and indexed at the bottom of the map. Some however, may opine that it is not necessary to plot the faults as they can be identified by the steep contour gradients in the interval map. The significance of mapping faults on *isopachs* may lead to deciding optimal drilling locations for the likely reservoir thickness to be encountered in the subsurface. The schematic diagram Fig. 2 illustrates the point. The predicted reservoir thickness encountered by drilling location varies across a fault zone as a function of fault heave, slip and fault type (normal/reverse). Depending on the ground location, vertical drilling in the fault zone may end up in missing or repeating a part of reservoir in case of normal and reverse faults respectively, a criterion commonly used to infer a fault from well data.

Paleo-Structural Analysis

Besides the thickness of the formations, the *interval maps* also reveal the earlier existing paleo slopes and structures at the time of deposition (Fig. 3). This provides vital clues for assessment of potential paleo-stricture prospects for hydrocarbon accumulations, discussed in detail in Chapter "Seismic Stratigraphy and Seismotectonics in Petroleum Exploration". Given an interval map between two horizons, the upper younger horizon can be flattened in the work station as a reference level resulting in thickness that shows the structural attitude of the older horizon. This signifies the paleo depositional surface of the older bed prior to deposition of the overlying younger horizon. This is known as 'paleo structural' analysis. For instance, referring to Fig. 3i, it can be seen that the older bed was an existing paleo-high while in Fig. 3ii, the existing flatbed became a high as seen in the current morphology, due to tectonics post deposition of the younger top bed. Similarly, in Fig. 3iii and iv, in the former the older bed was an earlier low, whereas, in latter case the older bed was deposited flat and deformed later on to become a low after deposition of the younger bed. The features are termed 'depositional or Paleo low' and 'structural low' respectively differentiate the two forms. Paleo-highs and lows have great geologic significance in assessing hydrocarbon prospectivity in exploration.

However, the analysis indicates paleo dips, azimuths and structures appropriately in a 'layer-cake' geologic setting where the reference, younger horizon is presumed to be deposited horizontally flat. The method may not offer meaningful paleo-structural information in case of depositional features like carbonate/clastic mounds, prograding sequences and erosional unconformities where the presumption does not

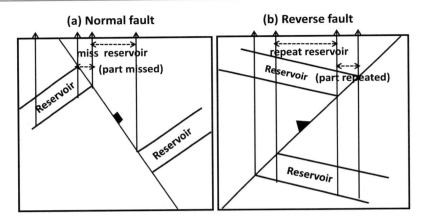

Fig. 2 Schematic diagram illustrating impact of fault planes on reservoir thickness which varies depending on the drilling location in the fault slip zone. The well would miss or repeat the reservoir thickness fully or partly depending on the fault type (**a**) normal fault and (**b**) reverse fault. Note the impact depends on the head and slip of the faults

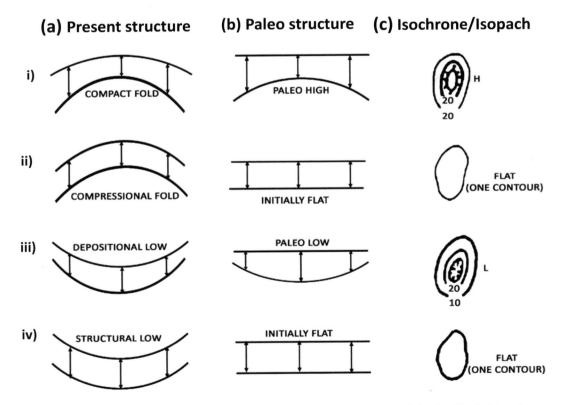

Fig. 3 Schematic illustrating paleo structural analysis done by flattening the top bed. (**a**) Structures as viewed presently, (**b**) structures existing at the time of deposition of the younger bed (Paleostructures) and (**c**) plan view of paleostructures contoured in time/depth interval maps (Isochrone/Isopach). Note the distinctions between (iii) Paleo low and (iv) structural low which have great geologic significance in hydrocarbon exploration

hold good. For instance, a flattening exercise with a carbonate mound may show its base, the depositional surface as a low which is misleading as carbonate mounds are known to occur on flattish or shoal surfaces.

Category II: Stratigraphic Interpretation (2D)

Stratigraphic interpretation is a high level synergistic technique to evaluate structural and stratigraphic details of subsurface geologic features, mostly prospective reservoirs. Usually, stratigraphic interpretation requires closer grid data and well data for qualitative estimate of rock properties for evaluating reservoirs for hydrocarbon production. It plays a much more important and exhaustive role in appraisal and development phases for reservoir delineation and characterization and is mostly carried out with high-resolution and high density seismic (3D/4D) data described in Chapter "Evaluation of High-resolution 3D and 4D Seismic Data".

Seismic Calibration

Stratigraphic interpretation starts with correlating the seismic response with the geologic information at the well, a process known as seismic-well calibration or seismic-well tie.

It is necessary to benchmark the seismic parameters with the log-measured rock parameters in the well so that the rock parameters determined from seismic can be predicted in areas away from the bore-hole. Seismic tie with well can be made in several ways that are briefly described below.

Overlaying Log Curves on Seismic

Overlaying log curves on seismic is the simplest quick-look way to match the seismic data with well geology. Workstations conveniently allow converting the log curves from depth to time domain using the sonic or well velocity and their overlays on seismic shows broadly the

correspondence between the seismic trace and the log curve at the well. The display highlights the major deviations, the excursions and incursions in the logs with respect to the corresponding seismic reflections which help identify major lithobreaks of the geologic formations. Some interpreters prefer to overlay the gamma-ray and resistivity curve and though they are excellent indicators of formation lithologies, there may be problem, for instance water saturated sands looking flat on resistivity curve would not correspond to seismic reflections. On the other hand, overlay of sonic and density logs, in particular, are more useful display as they represent the velocity and density of rocks which causes the seismic response. However, log overlays mostly identify the major mega formations and may miss finer details such as relatively thinner layers and their rock properties.

Continuous Velocity Logs (CVL)

Continuous velocity log (CVL) matching is very similar to overlaying log curves but shows the seismic-well correspondence a little more in detail for the target window. It is a plot of interval velocity (m/s) computed from sonic log transit time, which is transformed from the depth to time domain for match with seismic and displayed alongside the seismic segment for inferring lithology from the CVL formation velocity. It is a simple process and often provides useful correlation of seismic with geology at the well, especially for high-velocity limestone formations and where velocity measurements by check-shot or VSP data is not available to convert the log traces to depth (Fig. 4).

Seismic horizons calibrated at the well permit inferring geology away from the well from the variability in seismic waveforms characterized by amplitude, phase, frequency and polarity. It therefore is of utmost importance to have a rigorous seismic calibration. The essence of effective stratigraphic interpretation lies in the art of obtaining perfect ties of seismic with well log data, specifically with emphasis on target objectives, the reservoirs. This is commonly achieved

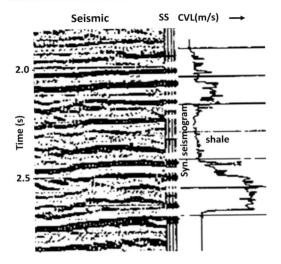

Fig. 4 Example of seismic calibration with CVL that shows excellent match. CVL calibration permits viewing of interval velocities which give details of the formations. The synthetic seismogram is also shown for confirming calibration. Note the central portion (shale section) displays flat and low value in the CVL and no reflections in the synthetic whereas the corresponding seismic section shows strong amplitudes. These are suspected multiples and are artefacts, warranting data reprocessing (*Image courtesy of ONGC, India*)

by synthetic seismograms and vertical seismic profiling (VSP).

Synthetic Seismogram

It is a simple forward modeling technique, most commonly used in calibration. Synthetic seismogram is a seismic trace computed by convolving an appropriate wavelet with the reflection coefficient (impedance) series determined from sonic and density logs at a well (Fig. 5). The computed synthetic trace with several of these repeated represent a short segment which is then spliced in the seismic for convenient viewing of matching between synthetic and real seismic at the well (Fig. 6). However, logs are sensitive to mud invasion and poor borehole conditions and require proper editing before computing the synthetic seismogram. The density log, in particular is highly susceptible to error due to borehole cavings and may be edited by the help of caliper log that measures the borehole dimensions. Similarly, the

sonic may require drift corrections (Chapter "Borehole Seismic Techniques") before computing the impedance log for preparing a synthetic seismogram.

Despite corrected logs and choice of suitable wavelet, the computed synthetic seismogram may still show mismatch with seismic (Fig. 6a). This is essentially due to matching a one dimensional response with the actual two dimensional response of field seismic trace. The specious inherent assumptions in computation of synthetic that lead to mismatch amongst others include.

- Reflection coefficients are computed for normal incidence only.
- Computing response for 1D vertical array of reflector points, in contrast to the concept of Fresnel zone width.
- Horizontal and plane reflecting layers in the subsurface.
- Wave propagation effects and noise not considered.
- The chosen wavelet chosen for computing the synthetic response is not time-varying.

Having said that, there can also be problems with seismic data, for resulting in poor match with synthetic. These reasons may be inadequate processing, particularly due to inadequacy in two dimensional migration of 2D data and complicated subsurface geology. Interpreter may find occasionally a poor match due to discrepancy in positioning of the subsurface seismic image vertically below the well position as could happen in highly dipping or faulted beds. Suspect for a mistie may also include the ground location of the well or the seismic trace, which would need the navigational check. In situations like this, the mis-ties can be investigated by shifting of the synthetic seismogram, by a few traces, in the vicinity of the well to check if it provides the desired match. The positioning problems are, however, more difficult to deal in strongly deviated wells and highly dipping horizons, especially with unmigrated/poorly migrated 2D seismic data of old vintages. Complicated nature of the subsurface geology can also be a cause for

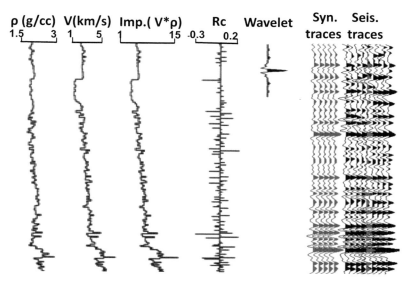

Fig. 5 Illustrates synthetic seismogram workflow. Reflection coefficients computed from impedance log (product of velocity and density) is convolved with a chosen wavelet to create synthetic seismogram and is posted alongside or spliced in seismic for comparison. Deviations would show the measure of calibration mismatch (*Image courtesy of Arcis Seismic Solutions, TGS, Calgary*)

Fig. 6 Examples of seismic calibration with a synthetic seismogram(SS) showing (**a**) poor match and (**b**) good match. Note the marked mismatch in amplitude and phase of reflections in (**a**) and also in the upper and bottom part of (**b**), but considered acceptable because of the good match at the target level. (*Image courtesy of ONGC, India*)

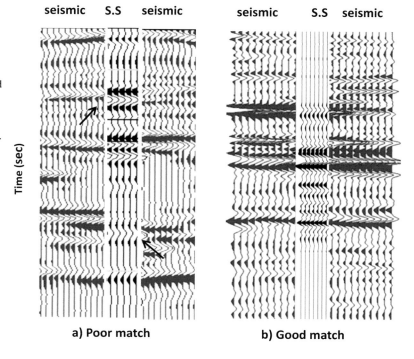

mismatch. Strong subsurface heterogeneity, proximity of fault zone generating considerable amount of noise and rapid facies change as often seen in fluvio-deltaic deposits and stratigraphic pinch out, can result in poor seismic image and make synthetic ties bothersome. However, recent advances in cartography and seismic processing have to a large extent have mitigated some of these problems.

A perfect synthetic seismogram seismic tie at all horizon levels may not be common. Generally it is considered a good match if an acceptable tie is obtained at the target and a few other prominent reflection horizons (Fig. 6b). The synthetic seismogram indicates the reflections and their signal attributes expected from the subsurface and matching with seismic also points to the quality of the processed seismic data. The mismatch, presence of more or absence of some reflections in the field seismic can be a matter of concern warranting investigation. Referring to Fig. 4 which shows an offshore seismic segment, synthetic and the CVL juxtaposed, may be considered as an example to demonstrate this. A thick shale section sandwiched between two limestone units, is indicated by a flat low sonic (CVL) and is corroborated by a zone of weak reflection in the synthetic. But the corresponding zone in in the seismic section evinces strong continuous reflection horizons. Clearly, these are not primary reflections and are spurious events

suspected as multiples, requiring reprocessing of data.

Vertical Seismic Profiling (VSP)

The VSP is generally the preferred tool for seismic calibration. VSP survey records seismic waves with a geophone in the well which is discussed in details in Chapter "Borehole Seismic Techniques". VSP besides providing accurate measurement of true vertical velocity, it records reflections from subsurface beds with a seismic source deployed on ground and provides better match with field seismic because both are similar systems (Fig. 7). The survey, however, increases drilling downtime and is sometimes compromised to minimize exploration cost.

With a proper seismic tie, stratigraphic interpretation permits proper delineation of prospects and prediction of structural and stratigraphic details such as depth, geometry and rock properties of a reservoir. Seismic data at this stage may need reprocessing or special processing with object-specific parameters to enhance resolution in relevant time windows for better clarity of the geologic target. Stratigraphic interpretation

Fig. 7 Display of seismic calibration with VSP corridor stack spliced in seismic segment. Note the good match as both belong to similar system and the reason for preferring VSP for seismic-well tie (After Balch et al. 1981)

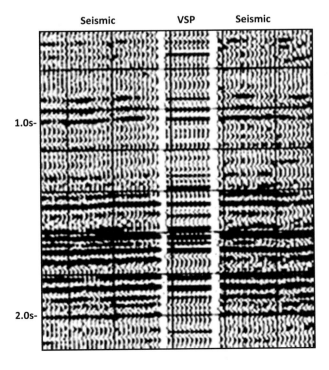

requires correlation of the target reflection horizons to be done more carefully and the continuity limited to the extent decided by the reflection character evinced in seismic. Phantoming and jump correlations may be avoided in general unless required for a specific purpose of mapping such as to show possible up-side in prospectivity. Reprocessing of data may have to be done taking into consideration the geologic factors and usually the aim to better the reflector continuity, may not be always justified by the geology. For instance in fluvio- deltaic environment where rapid facies changes are likely, in contrast to extended continuity, variations in reflection continuity and character of the horizon are preferred for mapping changes in facies which is the prime exploratory goal. Lateral variations in reflection character (waveform shape) of a horizon from that calibrated at the well strongly signifies changes in reservoir facies and properties and limits the areal extent of reservoir delineation (Fig. 8).

Depth Conversion

True average velocity estimation, though a hard job, is essential for reliable prediction of reservoir facies, depth, thickness, porosity and fluid content. Several techniques for velocity estimation from seismic are available for depth conversion and the appropriate choice depends on the interpreter's expertise and experience.

A simple and convenient way is to convert a time horizon map to depth map directly from the known velocity in the area, usually measured in a well. The map though shows correct spot depth at the well, it may show depth inaccuracies away from the well and consequently in structural disposition of the prospect. The map provides tentative depths useful in initial stage of exploration but can have significant impact in defining the prospect for choosing the subsequent drilling location. This is because, the spot-depth velocity function does not take into account the lateral variations in velocity, away from the well. It therefore necessitates the creation of a velocity model which considers lateral variations for depth conversion. Such velocity models can be built in several ways, contingent to seismic data grid density, availability of wells with check-shot and sonic velocity data and the accuracy required for the exercise at hand.

Velocity Modelling

In areas where no well velocity data exists, the velocity map can be prepared by contouring

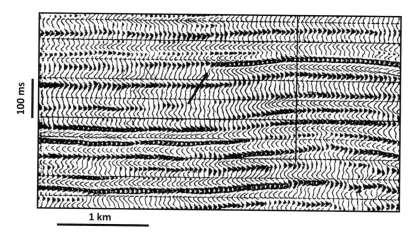

Fig. 8 Seismic segment showing change in reflection character of a horizon which indicates lateral changes in facies. Note the clear changes in amplitude and frequency (arrow marked) of the reflection which decides the limit of the horizon continuity. (*Image courtesy of ONGC, India*)

suitably-corrected RMS velocities and use it to convert time to depth. However, RMS ($\sim$NMO) velocities suffer from inherent drawbacks, such as dependence on the quality of seismic reflections, data grid noise and the processor's bias in picking the velocity during NMO velocity analysis. Due to inaccuracies, inherent in velocities obtained from seismic processing, the velocity maps may show localized highs and lows, known as 'bull's eye' which may be suitably edited and smoothened, constraining the geologic and structural trends in the area.

Where wells are available, but with no/meager well velocity or sonic data (as in old, underexplored blocks), a cross plot of datum corrected log depths of the target horizon against corresponding seismic times can provide an average velocity (overburden velocity) function for that horizon (Fig. 9). This is similar to a T-D curve, familiar to seismic interpreters, which is normally derived by measuring velocity in a well. Deviation from the regressive straight line fit in the cross plot signifies lateral variations in velocities which may be used limited to that part of the area. However, the method requires unambiguous correlations of horizons in the logs with the corresponding times in seismic which may not be always feasible. In cases where velocity data from several wells, spread over the area exists, the velocity values calculated from the wells for the target horizons can be contoured to yield a velocity map for the horizon(s) and used for preparing depth maps.

However, a stringent velocity modeling using well and seismic velocity, besides providing more reliable depth conversion has added advantages. The model can be evaluated numerically and geologically for reasonableness and as it enables the use of velocity information from both seismic and wells, it provides scopes for critical review and quality control (Etris et al. 2002). Velocity modeling may use simple average velocity for single layer, interval velocity for multi-layers, or instantaneous velocity for velocity variation in layers with depth (Etris et al. 2002).

Depth conversion of target horizons by using a constant average velocity assumes the overburden as one single formation and lacks providing

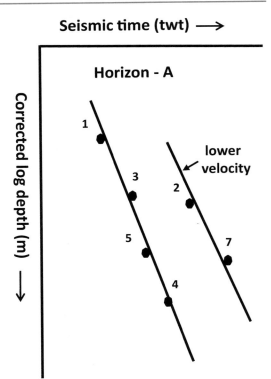

Fig. 9 Graphic illustrating estimation of average overburden velocity by cross-plot method in areas where several wells are available but without velocity/sonic data. Datum corrected log depths of horizon (A) and the corresponding seismic times at the wells are cross-plotted and the best fitted straight line provides the average velocity of the horizon with depth. Shift from the gradient (arrow) indicates a slower velocity signifying lateral velocity variation in the area. (Courtesy, ONGC, India)

velocity information of horizons above though it is an easy and fast way, often practiced to convert depth of the target horizons (Fig. 10a). Another approach to velocity modelling may be based on interval velocities (Fig. 10b). A workstation based technique called 'stripping method' is often employed to estimate the velocity function. The interval velocity at the well versus depth curve is analyzed for building velocity models for each horizon from the shallowest to the deepest. With the first horizon depth known, depth to next and successive horizons are estimated by adding respective individual horizon thicknesses (isopachs) calculated from interval velocity. The first horizon is usually the shallowest marker converted to depth by tying the log depth at the

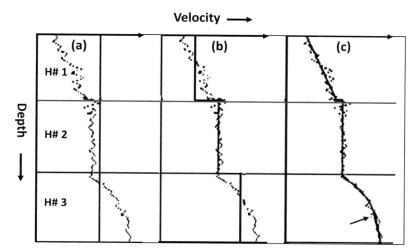

Fig. 10 Graphic illustrating velocity modelling for depth conversion using (**a**) average velocity, (**b**) interval velocity and (**c**) instantaneous velocity. (**a**) Note the average velocity marked by the black vertical line assuming the overburden as one single formation. (**b**) The interval velocity is constant in each horizon and provides

information about above horizons. (**c**) The instantaneous velocity changes with depth within the horizons and provides more details about velocity behavior in the layer by its gradient. More details can be obtained when the gradient is modelled parabolic instead of linear (after Etris et al. 2002)

well. The computed depths at the successive horizons are checked for match at the wells, where available. In case of mismatches, the velocity field may be refined till satisfactory match is achieved at all the horizons.

The interval velocities used in the process are the mean interval velocities for the formations. For better accuracy, models can be built by using instantaneous velocities (Fig. 10c). The instantaneous velocity at any depth 'Z' is determined from the gradient by the equation $Vz = V_o + kZ$, where 'V_o' refers to the velocity at a reference point, usually the seismic reference, and 'k' is the gradient. This allows computation of interval velocities for beds of small thickness so that depths can be predicted layer-by-layer. This can be further improved to provide more accurate results by adding a curvilinear fit instead of a linear fit in the model (Fig. 10c) However, in relatively low resolution and low-density 2D seismic data, the technique may suffer from varying orders of inaccuracies and may not adequately account for lateral velocity variations due to poor control on extrapolation of data beyond the wells.

Velocity maps have an advantage in that the trends and variations can be viewed in the area *vis-a vis* the structural features in the time map. Significant anomalies observed in velocity trends and not consistent with time map may be analyzed for the reasons of discordance and supported by geologic trends for having confidence in predicting the velocity model and consequently the depth conversions.

Case Example

A simple but important practical clue to variation in lateral velocity can be realized by simple method of displaying the drilled depth values of a horizon at the wells on the time map (Fig. 11a). Discrepancy in relative variance of depths with corresponding seismic times at the wells reveals presence of lateral velocity variation in the area. As an example of a real case, consider the three pairs of time-depth values, the 1340 ms (1470 m), 1440 ms (1600 m) and 1500 ms (1510 m) in Fig. 11a. The inconsistency in the pattern of variance of the last pair with respect to the other two, can be clearly seen. This indicates presence of lateral velocity variation, lower velocity in this

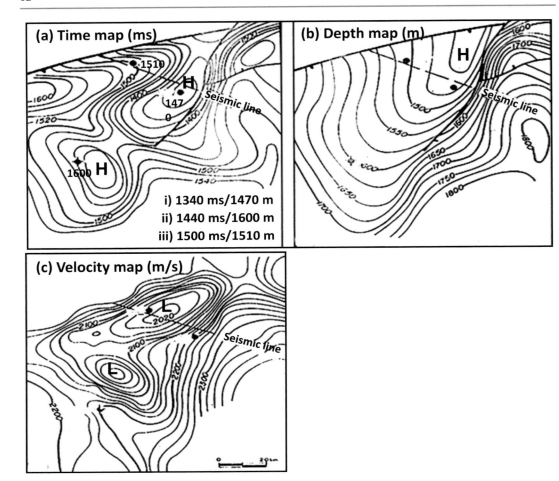

Fig. 11 Illustrates method of well depths plotted on time map to check time-depth relations that could signify lateral velocity variation. Of the three sets of time and depth shown at bottom of map (**a**), notice the inconsistency in the third set. The depth increase is marginal corresponding to much higher time value which clearly indicates lateral velocity variation. Velocity map prepared from 2D seismic confirmed the slower velocity in this part. Note the impact, the dissimilarity between the time and depth maps 12a and 12b (after Nanda et al. 2008)

case and the impact of velocity variance was significantly large on the depth map (Fig. 11b). A velocity map, subsequently prepared from 2D stack velocities, corroborated the existence of the low velocity in this part (Fig. 11c) and explicably, the reason for the low velocity is due to presence of a channel cut filled with shale above the target horizon (Nanda et al. 2008).

In such cases, where the actual depth gradient at the wells is opposite to that of time gradient, the presence of severe velocity variation in the area can be a forewarning for the interpreter to be extra cautious about depth prediction. As stated earlier, such anomalous velocity trends linked to geological reasons may be analyzed to suitably incorporate in editing the trends in the velocity models. The velocity problems, however, are much better mitigated in 3D seismic because of higher density of sampling and better resolution of data, though velocity maps prepared from seismics velocity volumes may still need supervision.

Interval Velocity for Evaluation of Rock Properties

Interval velocities computed from seismic RMS velocity, and duly constrained by sonic velocity, are often used to predict qualitatively reservoir facies, porosity and fluid contents. A lowering of velocity is usually considered as an indication of higher porosity and a severe velocity lowering may indicate gas in reservoir in some cases where the change is presumed not due to lithology. However, interval velocities are highly sensitive to ambiguities in stacking velocities, especially when computed for smaller intervals and can be misleading in case of thin reservoirs. In relatively low resolution and low density seismic 2D data, velocity prediction may suffer from varying degrees of inaccuracies depending on quality and grid of data. Fortuitously, in early stage of exploration, high accuracies in depth predictions may not be generally warranted.

Seismic Structure Maps

Structure maps, besides being the key deciding factor for guiding locations for drilling successful wells, also provide prime inputs for volumetric estimate of in-place hydrocarbons. As such, accurate seismic depth maps play a vital role in exploration and development of fields but its prediction accuracy can be a challenge to the skill and experience of an interpreter. The precision of the map depends on meticulous picking and correlation of the appropriate seismic phase in the initial stage and estimating the accurate velocity function at that level. It is also to be ensured that the contouring, including that of the faults, be geologically appropriate, especially for stratigraphic features and fault associated traps. As mentioned, precise velocity estimation is by far the most crucial factor for creating accurate depth maps and the prediction errors in structural configuration of depths and thickness of the reservoir can lead to disappointing drilling results and severe consequences in estimating the in-place hydrocarbon reserves. Economic viability of a field can change significantly if an appraisal well meets the actual pay deeper and ends up in a water well. Empirically, a departure up to 10% in 2D data between predicted and actual depths may be considered tolerable at moderate depths (~ 2000 m). The prediction accuracy decreases with depth, as errors in seismic derived velocity usually shows larger deviations at deeper depths.

Discovery in a well shifts the project from the initial phase of exploration to the next phase for delineation of the reservoir. Invariably, the old seismic data is revisited for fresh interpretation and often suggested for reprocessing or special processing of data. Sometimes new data with closer grid may be necessary to be acquired. With better quality new data and geologic information gained from the drilled well(s), reinterpretation produces a set of new, more precise structure maps to predict depths and geometry of the reservoir. Contouring, depending on geologic leads from wells, may require explicit representation of subsurface geology, especially for the stratigraphic features such as channel sands, delta lobes, bar sands etc. It calls for special contouring skill to properly portray the stratigraphic features and needs human intervention in computer-based contouring.

Contouring Example

An example of hand-made contour vis-a-vis computer-made contour map is shown (Fig. 12). The realistic geologic model shows a barrier bar cut through by a river channel with the lagoonal and back swamp area inland (Fig. 12a). The data points indicate the associated sand facies and thicknesses. The contour maps prepared manually are shown in Fig. 12b and by the computer using co-kriging in Fig. 12c. The maps, however, are remarkably different. The repeat of the 20 m contour in the manually prepared map (Fig. 12b) signifies decreasing sand thickness to north as is to be expected from the geologic model of the swampy lagoonal area inland. Whereas in contrast, the computer-generated map indicates thickening of sands to the north (Fig. 12c), contradicting the geologic model (Fig. 12a). This is because the computer failed to repeat the 20' contour value. The contour trends of sand thickness are also dissimilar which can lead to major discrepancies in terms of reserves. The computer

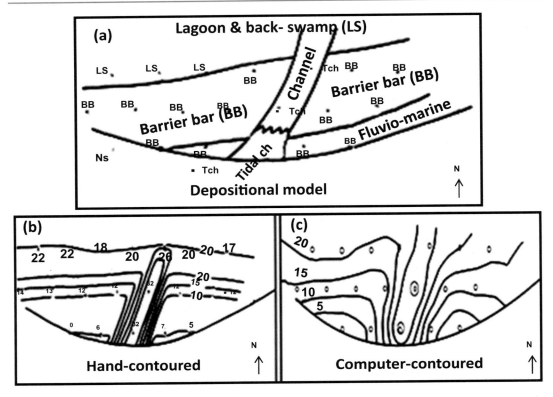

Fig. 12 Showing hand-contoured and machine-contoured maps of a stratigraphic prospect. (**a**) Depositional model (**b**) hand contoured map of sand thickness and (**c**) computer generated map. Note the significant difference in trend of contours. The repetition of the 20′ contour in (**b**) indicates thinning of sands to north in the swampy lagoon area whereas the computer generated map (**c**) indicates increase of sands in this part, contradicting the geologic model because computer failed to repeat the 20′ contour value. It also could not contour the sand thickness parallel to coast as it ought to be by a barrier bar. Also note (**b**) the contours depicting the barrier bar appropriately with steep slope inland and flat coastward

prepared map does not show sand-thickness contours parallel to coast as it should to be in a barrier bar feature. The hand drawn contour map (Fig. 12b), on the other hand depicts the barrier bar appropriately with steep slope inland and flat coastward. The difference is because the human imagination incorporates geologic information on structural and stratigraphic analogs in mapping while co-kriging only interpolates the data mathematically and mechanically regardless of the geologic feature.

As mentioned earlier, detailed reservoir maps are normally prepared with small contour intervals (10 ms) and small scales (1:10,000/20,000) to signify the reservoir details better. It is also important that all cartographic details are correctly displayed to avoid any uncertainty in location

(ground-position) of wells and seismic for ties. The revised seismic maps also need to be supported by the known hydrocarbon distribution and fluid contacts at wells for better reliability. For instance, a map prepared at the top of an oil-pay is normally not supposed to indicate structural disposition lower than the established oil/gas -water contact. If, however, such an anomalous situation is confirmed after due checks, it will have major implication on the significance of distribution of hydrocarbon pays in the prospect. This may indicate another discrete isolated oil/gas pool, and with a possibility of many such multiple pays which can have great consequences in the development of the prospect. Faults are an important component of structural maps, and play a crucial role in the distribution and accumulation of hydrocarbons and they need be carefully mapped and cautiously evaluated (Chapter "Seismic Stratigraphy and Seismo-tectonics in Petroleum Exploration").

Seismogeological Section

Seismogeological transect is an effective composite display of well-log correlations superimposed on seismic correlation (Fig. 13). These are convenient and preferred ways of presentation to understand and assimilate the stratigraphic details across the profile with its log and related seismic response. Logs are often correlated based on lithostratigraphic correlation and may vary from the seismic chronostratigraphic correlation, at times the trends crossing each other. The composite seismo-geological display provides an opportunity to check log and seismic correlations and discordance, if noticed, necessitates a reconciliation by modifying correlations in one or in both the data sets. However, reliable seismic horizon correlation representing a stratal surface can lead in most cases to guide modification in the log correlations. For instance the log characteristics of two sand units in the two wells may be extremely similar yet may be discrete belonging to different ages. Log responses linked

to similar genesis of geological depositional setting may show alike motifs but may not be continuous from one to the other well. The log markers may not exactly match seismic markers but the correspondence is usually considered satisfactory if the trends are parallel.

Category III: Seismic Stratigraphy Interpretation

Seismic stratigraphy is a geologic approach to interpret regional stratigraphy from seismic. It is a powerful technique, especially suitable for less explored or virgin basins with no or sparse well data. Even a few long routinely processed regional seismic lines can help to capture information on the evolutionary history of the basin and broadly identify and assess its hydrocarbon potential. Hydrocarbon generation, migration, and accumulation potentials in the basin in general and the favorable structural and stratigraphic plays in particular, can be assessed and the resources prognosticated for deciding future exploration strategy in the basin.

The seismic stratigraphy method is based on analysis of reflection patterns of stratal surfaces. A strata may be considered as a continuous deposition within a geologic time and its top, the

Fig. 13 Displaying a composite profile of seismic segment with well logs to comprehend better the seismic response of the changes n lateral facies seen in the well logs. Note the change in reflection character of shelfal limestone to that of sandstone at shelf break evinced clearly in seismic (indicated by arrows). (Modified after Galloway et al. 1977)

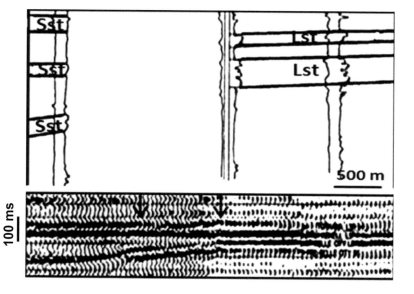

stratal surface signifies a period of no deposition or a change in depositional regime, thus forming time-lines through the geologic process (Mitchum 1977). The reflections from impedance contrasts at the interfaces tend to parallel the stratal surfaces and seismic correlations are usually considered chronostratigraphic though they are strictly not so. Chronostratigraphic correlation helps study of sequences of rock units deposited during a given span of geologic time. Commonly, the seismic image portrays the subsurface geology adequately in simple geologic settings. The reflections and their patterns can thus be taken as fair representatives of geometry of the stratal surfaces (reflectors) and their depositional morphology. Seismic chronostratigraphic correlation, however, can be at variance and cut across other types of correlations such as lithostratigraphic or biostratigraphic correlation which may have similar facies but belonging to different geologic times. Facies deposited during different interval of geologic time are known as diachronous and their correlation is termed time transgressive. Depositional facies involve the sediments carried by rivers and deposited on land or at sea influenced by changes in sea level. We shall mostly restrict to marine deposits and briefly state using the standard terminologies of seismic stratigraphic framework (Mitchum and Vail 1977a,b).

Seismic stratigraphy framework consists of three parts:

(i)   Seismic sequence analysis
(ii) Seismic facies analysis and
(iii) Relative sea level change analysis.

(i) **Seismic sequence analysis**

Seismic sequence analysis is about identifying the depositional sequences. A sequence is defined by unconformities or correlative conformities at top and bottom of a rock unit and is recognized from discordance of reflections as seen by lateral termination in seismic sections. An unconformity is a surface of nondeposition, separating older from younger rocks and a correlatable conformity is a surface which can be correlated to an unconformity elsewhere in the basin. Reflection terminations are of two major types; younger horizons terminating against older at the base, termed as baselaps and older horizons against younger at top, known as toplaps (Fig. 14). The baselaps and top laps each are sub-classified to two kinds. The discordant relation of each type of reflection is outlined, illustrated by sketch and seismic example which are self-explanatory to facilitate easy understanding of their geologic significance.

Onlaps, coastal and marine
Onlap, the discordant relation at the base can be coastal onlap or marine onlap. Landward

Fig. 14 Schematic showing types of seismic reflection termination patterns that identifies depositional sequences. (Top) Reflections terminating against older formation are baselaps and (bottom) terminating against younger formations at the top as toplap sequences. (After Vail et al. 1977a, b)

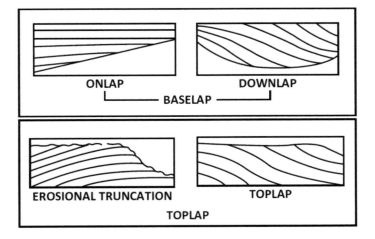

building of nonmarine coastal sediments on shelf is known as coastal onlap and building marine sediments on basin slope is known as marine onlap (Fig. 15). Onlaps generally are indicative of relative sea level rise.

Downlap

Similar to onlap, it is also a base discordant relation and indicates lateral extent of stratal deposition, basin ward (Fig. 16). The down- dip and the gradient of the strata indicate the transport direction and rate of sediment supply. Large and fast sediment dumps are likely to show steep gradients.

Toplap

This is a top discordant relation and shows termination of an older strata at the top against younger unit (Fig. 17). Coastal top laps represent

non-deposition and/or mild erosion above wave base in a still-stand sea level. Toplaps also can be seen in deep marine depositions.

Truncation

It is another top discordant relation, which shows strong angularity with overlying younger strata and indicates erosional unconformity. The erosional surfaces due to weathering are common on land but also occur in offshore created by huge submarine canyon cuts (Fig. 18).

(ii) **Seismic facies analysis**

Seismic facies analysis is the most important component of seismic stratigraphy. Geologic facies vary in lithology and rock properties depending on depositional environment and sedimentary process and have individual

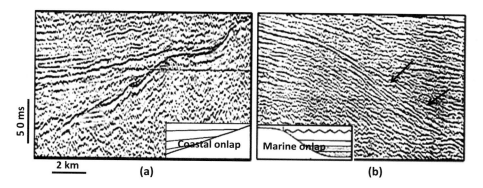

Fig. 15 Displaying seismic images of onlaps, base discordant relation with older formation with schematic (insets). (**a**) Coastal onlaps back-stepping on the coastal part and (**b**) marine onlaps back-stepping against basin shelf slope. (Image courtesy of ONGC, India)

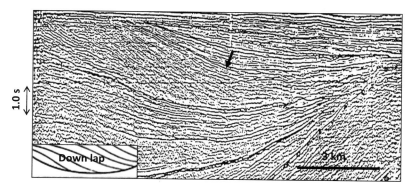

Fig. 16 Seismic segment displaying down lap sequence (arrow), a base-discordant relation with older formation with schematic (inset). Also note the base-discordant onlap sequence below it (*Image courtesy of ONGC, India*)

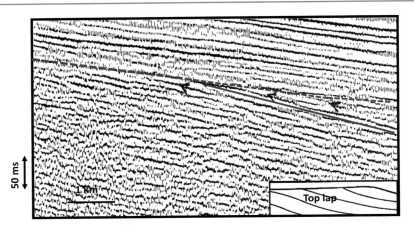

Fig. 17 Showing seismic example of top lap sequence, a top discordant relation with younger formation (arrow marked) with schematic (inset). (*courtesy, ONGC, India*)

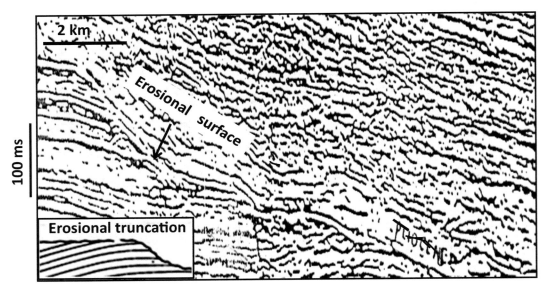

Fig. 18 Seismic segment showing a top discordant erosional truncation with schematic (inset). The erosional surface is created by a huge Pliocene canyon cut in deep offshore (*Image courtesy of ONGC, India*)

characteristic signatures imprinted in seismic images. Seismic facies analysis deals with analysis of seismic internal reflection patterns with their attendant characteristics within a sequence to interpret depositional environments and related lithofacies. Facies analysis of a sequence considers reflection characteristics which include continuity, amplitude and frequencies, internal reflection configurations, interval velocity and external forms.

Reflection continuity and amplitude

Reflection continuity over large areas with consistent amplitude is an indication of wide spread uniform depositional environment, implying mostly marine facies. Discontinuous reflections with variable amplitudes in an area, on the other hand, imply frequent lateral changes in facies indicating continental to deltaic environment. High amplitude continuous reflections are usually suggestive of sandstone or limestone lithology. Zones of weak amplitudes spread over a wide area are likely to represent monotonous lithology, usually inferred as marine shales though it could occasionally be large continental alluvial plain sands deposited under fluvial environment.

Frequencies

Frequencies often indicate age of sequences and facies changes (Chapter "Seismic Reflection Principles—Basics"). Low-frequency reflections at deeper and high-frequency dominated reflections at shallower part of seismic records provide clear clues to confirm identification of sequences by indicating unconformities and also their relative geologic ages e.g., older Mesozoic rocks top lapping younger Cenozoic rocks. Frequencies also indicate relative bed thicknesses, higher frequencies indicating thinner beds.

Internal reflection configuration

Reflection patterns within a sequence, considered as a replica of internal geologic stratal configuration, are important clues for inferring depositional environments. Depositional environment though depends on several factors, it mainly involves, the type of transport agency, the amount (load) of sediments it carries and the place where they are deposited that matters the most. Energy refers to the force of the agency which carries the sediments, principally the rivers and to lesser degree waves and tides. Low energy deposits are related to finer clastics like clay and high energy to coarser clastics like sands. Deposition of sediments is controlled essentially by two factors: (i) the energy of the agent transporting the sediments, and (ii) the accommodation space available at the place for deposition. The sedimentary depositional patterns are controlled by a balance between the influence of the above two factors, the rate of supply of sediments, often referred as influx and rate of basin subsidence allowing space to deposit. This is discussed more under relative sea level rise analysis later in the chapter. Some of the common seismic internal configurations with their geologic significance are briefly mentioned.

Parallel and divergent reflection patterns

Parallel to sub-parallel pattern of reflections within a sequence suggests a balance between even rate of sediment deposition on a uniformly subsiding or nearly stable basin shelf. Reflections with divergent patterns, on the other hand, imply lateral variation in rate of deposition on a progressively tilting surface (Fig. 19).

Prograding Clinoforms Patterns

Progradation is an out-building of sediment deposits, caused by relatively higher rate of sediment supply and comparatively lower rate of subsidence. The low rate of subsidence does not provide enough accommodation space for the sediments to deposit and results in forcing it to out-build. Prograding clinoforms are sloping depositional strata with their forms influenced by transporting agencies, the amount of sediments they carry and their energy and are excellent indicators of unconformities, relative sea level changes, paleo-bathymetry (at time of deposition) and depositional energy. The top and base of clinoforms offer clues to paleo-bathymetry, the height of standing water column in which sediments were deposited. This is helpful in distinguishing prograding deltas, typically in shallow waters from those progradations occurring in deeper waters. Several types of clinoforms are formed depending on their geometry (Fig. 20); however, the three most significant and common clinoforms, the sigmoidal, oblique and shingled are described.

- Sigmoidal Clinoforms are recognized by vertical building, known as aggradation of depositional surfaces (Fig. 21a). Strong aggradations signify high rate of sediment supply and rapid rise of relative sea level which provides large accommodation space for vertical stack of strata. Sigmoidal clinoforms are suggestive of deep-water and low-energy depositional environment.
- Oblique Clinoforms are characterized by distinct toplap and downlap terminations. The progradation is termed tangential oblique (Fig. 21b) or parallel oblique (Fig. 22) depending on the steepness of the angle made at the lower boundary. The patterns suggest high sediment supply with slow to no basin subsidence in a relative still- stand sea level and a high energy depositional environment.

Parallel/sub parallel reflections

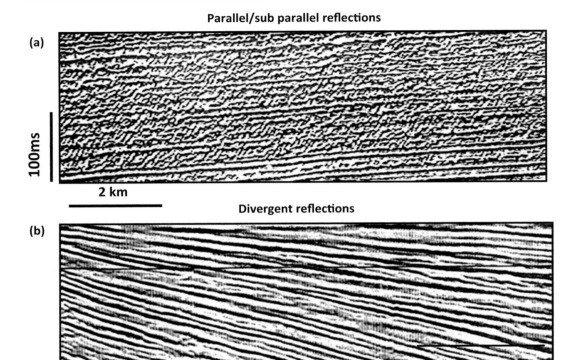

(a)

100ms

2 km

Divergent reflections

(b)

Fig. 19 Seismic example showing internal reflection configurations of seismic sequences. (a) Parallel/sub parallel reflections suggestive of steady rate of deposition on a flat and uniformly subsiding shelf. The mild dips seen are due to post depositional tilt. (b) Divergent reflections suggest uniform depositional rate in the basin which is subsiding slowly and progressively with mild tilt. (*Image courtesy ONGC, India*)

- Shingled Clinoforms are similar to parallel oblique (Fig. 23) patterns except that the clinoforms are much gentler and with a hint of an apparent toplap and downlap at the base. The low gradient of the prograding surfaces are suggestive of relatively less supply of sediment deposited on shallow stable platforms in shallow marine to deltaic high-energy environment.

Hummocky, Chaotic and Reflection-free Patterns

Hummocky patterns are often associated with low energy facies such as interdeltaic and pro-delta shale. Chaotic pattern of reflection on the other hand, is often linked to high energy deposits such as channel fills, fans, reefs, turbidites, etc.,. Reflection free zones are typically related to complex tectonized zones, core of salt or mud diapirs and over-pressured sections.

Interval velocity

As stated earlier, interval velocity is an important indicator of rock properties that can be calculated from sonic or average velocity measured in a well or estimated from seismic stacking velocity. Stacking velocities, despite their limitations, can still be extremely useful, if analyzed cautiously for prediction of lithology (Chapter "Seismic Reflection Principles—Basics"). However, difficulty arises as more than one type of lithology often has overlapping velocity ranges. Analysis of depositional environment by seismic stratigraphy approach combined with velocity can lead to reliable prediction of lithology and its changes within a sequence.

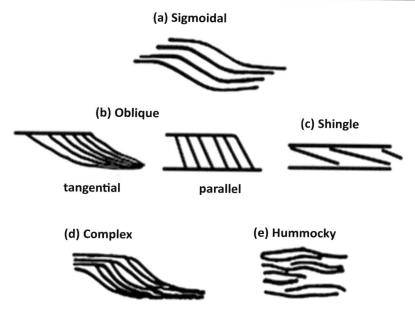

Fig. 20 Schematic diagram showing different progradational clinoforms, their morphology indicating sediment influx, sea level change and depositional energy. (**a**) Sigmoidal indicates large influx, rise in sea level and low depositional energy, (**b**) large influx, stand-still sea and high energy, (**c**) moderate supply of sediments, stand-still sea and moderate energy, (**d**) large supply of sediments, stand-still sea alternating with rise in level, high energy and (**e**) relatively low influx, sea level rise and mixed energy. (Figure after Mitchum et al. 1977a,b)

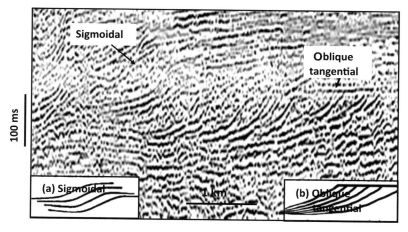

Fig. 21 Seismic segment showing sigmoidal and oblique tangential progradations with schematic schematics inset, (**a**) and (**b**). Sigmoidal progradations are characterized by vertical build ups (aggradations) suggesting high influx, rising sea level and low energy deep-water environment. Oblique tangential clinoforms with toplap suggest high influx, high energy deposits in still-stand sea level (courtesy, ONGC, India)

External forms

External form is the study of geometry of a seismic sequence in a profile as well as in plan-view (map) and is extremely helpful for geological support of seismic interpreted depositional features. For instance, a channel system interpreted from seismic when mapped in plan-view should be upheld by the trend in the down

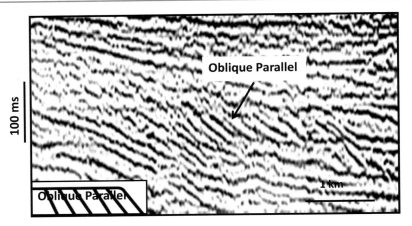

100 ms

Oblique Parallel

Oblique Parallel

1 km

Fig. 22 Seismic example showing 'oblique parallel' clinoforms with schematic (inset). Toplap Indicates relatively high influx, high energy and still-stand sea level. Note the relatively steeper stratal dips suggestive of rapid deposits of large amount of sediments (*Image courtesy, ONGC, India*)

Fig. 23 Seismic example of shingled progradation with the schematic (inset). Shingle clinoforms suggests low influx, high energy, stand-still sea level, typical of deltaic to shallow marine environment. Note the subtle suggestions of toplap marked by arrows (*Image courtesy, ONGC, India*)

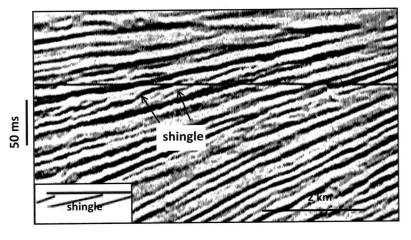

50 ms

shingle

shingle

2 km

dip direction and transverse to the paleo coast. The discordance patterns of reflection, internal configuration, formation velocity, and the external form of a sequence analyzed together, offers dependable information on depositional energy and associated facies of the subsurface geology. Some of the more commonly occurring external forms and easy to figure out on seismic sections and maps include wedge outs, fans, lenses, mounds and trough-fills.

Wedge outs show an external form of the sequence tapering land ward in seismic profile and its depositional environment and associated facies can be inferred from analyzing internal configuration of reflections. Fans, lenses and mounds have typical external forms seen in seismic with discrete amplitude anomalies and limited areal extent. These features are usually linked to high energy facies. The features interpreted from seismic profiles mapped in planview should be supported geologically by their trend and disposition. For instance, mapped fan deposit prospect is likely to show shapes elongated and trending in dip direction along the direction of sedimentary influx. Whereas in contrast, mounds would show nearly circular in shape, a three dimensional feature without specific trends. Nevertheless, it may be mentioned that the seismic responses discussed above are paleo-depositional trends which may be disturbed by subsequent tectonics and would need paleo depositional analysis to corroborate geological history.

Troughs, canyons and channels with their varied types of fills, offer interesting studies of reflection patterns and are good indicators of basin subsidence, depositional energy and facies. A schematic diagram of different type of trough fills is shown in Fig. 24. Divergent reflection fills in a trough (Fig. 25) with good continuity and amplitude denote low energy marine facies with gradual subsidence of the trough. In contrast, fairly continuous and near-parallel high amplitude reflections of a canyon-cut fill indicate low energy finer sediments with a fairly stable canyon without subsidence (Fig. 26). Yet another example of canyon cut and fill shows Irregular and chaotic trough fills with variable amplitudes indicating depositional facies of variable energy such as debris flows and gravity dumps of finer clastic deposits (Fig. 27).Trough- fills showing discrete erratic pattern of high amplitudes with mound forms signify high energy deposits, likely to be tur-bidites, channel cut and fill sand complexes (Fig. 28). The interesting examples of seismic images highlight the potency of seismic facies analysis technique where external form of a sequence and its internal reflection configuration is analyzed together, provide useful information to evaluate likely hydrocarbon prospects.

(iii) **Relative sea level change (RSL) analysis**
One of the key components of a depositional environment is the sea and its rise and fall with time that controls the shoreline and consequently the nonmarine, deltaic and marine depositional environments. Sea level changes on a global scale are known as eustatic changes and used for glacial and global geochronology studies. An apparent rise or fall of sea level with respect to land surface in a regional way is known as the relative change of sea level (Mitchum 7) which happens due to change in amount of water in the sea or due to subsidence or upheaval of land. Relative changes in sea level (*RSL*), controls the depositional processes and their patterns and is the key to understand and evaluate prospects for hydrocarbon. The depositional process involves amount of sediment influx, type of energy, and position of the shoreline and accommodation space available and is illustrated in Fig. 29. The process eventually influences the type of sedimentary deposits, facies and their thickness depending on the place where they were deposited.

Fig. 24 Schematic diagram illustrating external forms of troughs and their fills with varying internal reflection configurations which serves as a good indicator of subsidence, load of sedimentary influx, depositional energy and associated facies. (After Mitchum et al. 1977a,b)

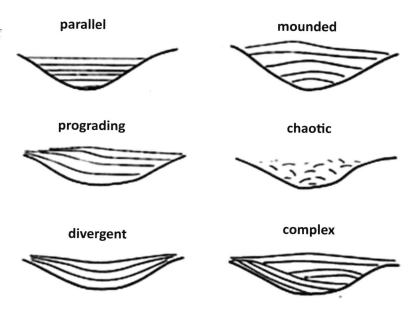

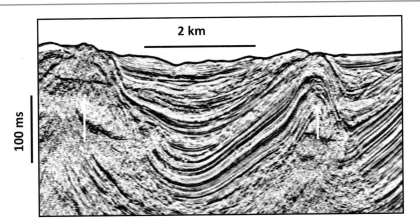

Fig. 25 Seismic image showing a trough and its fill with divergent reflection configuration which signifies continuing sedimentation with a gradual sinking of the trough, perhaps due to the continuing rise of flank diapirs shown by arrows. (*Image: courtesy ONGC, India*)

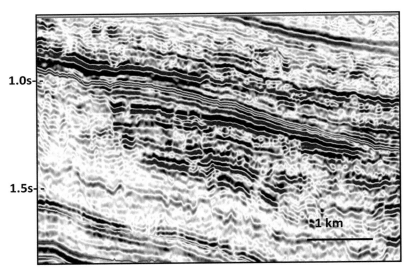

Fig. 26 Displays seismic image of a trough, a canyon cut and its fill. The cut and fill complex showing high amplitude, near-parallel and fairly continuous internal reflections indicate low energy finer clastics in deep water environment. Note the near-flat trough-fills are punctuated signifying lateral switching of the canyon, stable without any subsidence The dip seen is due to post-depositional tilt. (*Image: courtesy ONGC, India*)

Important inferences of relative sea level changes and associated facies can be made from seismic sequence and facies analysis. A rise in relative sea level is indicated by progressively shifting of coastal onlap sequences landwards, whereas a fall in sea level is indicated by shifting of the onlap sequences seaward, also termed as offlap (Fig. 30a). Fall in sea level can also be distinctly inferred from seismic where prograding clinoforms are seen shifting progressively downward towards basin (Fig. 30b). The seaward shift of shoreline with prograding regressive facies during a fall in sea level is known as '*forced regression*' (Fig. 30b), whereas regressive facies during rise in sea level which is more common, is known as normal regression (Fig. 30c).

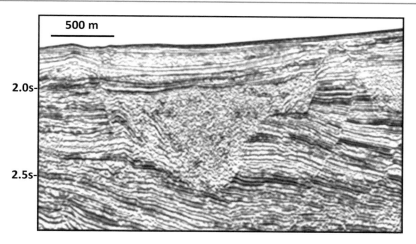

Fig. 27 Another seismic example showing internal reflection configuration of a canyon cut and fill in deep waters. The irregular chaotic reflection pattern with variable amplitudes suggests depositional facies of variable energy such as debris flows and gravity dumps of finer clastic deposits (*Image courtesy of ONGC, India*)

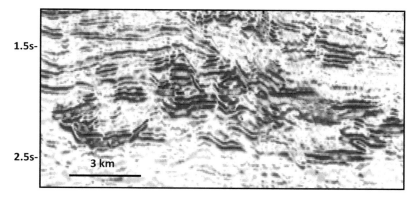

Fig. 28 Seismic image of a trough fill with chaotic and mounded internal reflection configuration. The strong amplitude and random mounded reflection patterns suggest high energy coarser clastics deposits in submarine river channel complex

Rise or a fall in sea level does not always determine the associated facies as transgressive or regressive. The marine transgression and regression is defined by shifting of shoreline, landward or basinward and not necessarily controlled only by sea level rise or fall. Shoreline is defined as the divide between marine and non-marine facies, and the shore facies are characterized by littoral deposits, defined as the sediments deposited between high and low tides. If the littoral deposits, linked to shore line, extend landward it is a marine transgression and if the shift is seaward, it is regression of sea. Mapping shoreline is an important aspect for locating potential hydrocarbon prospects because the high-energy reservoir facies of exploration interest mostly occur on or close to shorelines, proximal to hydrocarbon source located in the basin. However, mapping paleo shoreline from seismic without well information may be a difficult task.

It is interesting to note that shoreline shifts regardless of rise and fall of the sea level can be controlled by the amount of sedimentary influx

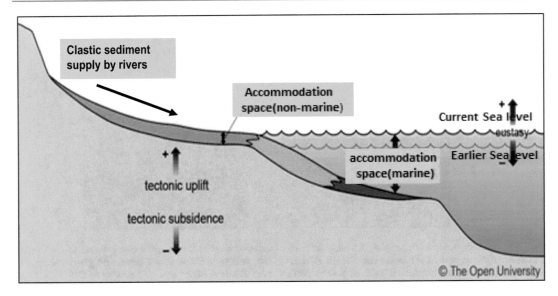

Fig. 29 Diagram illustrating depositional process in marine environment. Sediment influx direction, carrying agency (river), sea level changes due to uplifts and subsidence and the resultant accommodation space created for sediment deposition are shown. Note the accommodation space also provided by the river (non-marine) in low stand sea level. (Image courtesy: The open university, UK)

Fig. 30 Schematic graphic illustrating regressive facies associated with relative sea level changes. (**a**) Seaward shift of onlaps (offlap) with fall in sea level, (**b**) showing similar geometry of clinoforms shifting seaward under sea level fall, known as 'forced regression' and (**c**) regressive facies during sea level rise and known as 'normal regression'. (Modified after Vail et al. 1997)

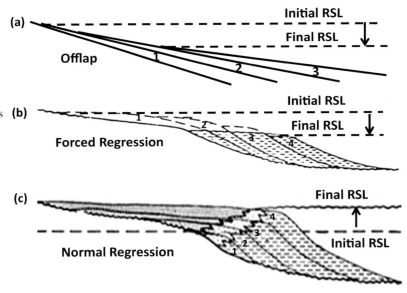

brought by the carrying river. For instance, during a sea level rise, transgressive, regressive or stationary shoreline facies can occur (Fig. 31), controlled by the balance between the accommodation space provided by relative sea level changes (including subsidence) and amount of influx supply. Consequently, during transgressive sea, a regressive facies, such as a prograding deltaic sequences occur which is common worldwide.

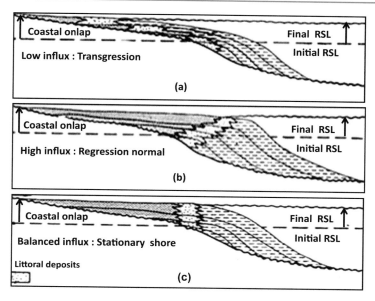

Fig. 31 Sketches illustrating shore line shifts and associated facies during relative sea level rise. The shifting of shore line, land or sea ward or remaining stationary depends on amount of sediment supply. (**a**) Landward shift of shore line(littoral deposits) results in transgressive facies with low influx (**b**) basinward shift yields regressive facies when influx is high and (**c**) stationary shoreline with rate of influx balanced with rise in sea level. Note in all the three cases, the sequences are coastal onlaps occurring with sea level rise. (After Vail et al. 1977a, b)

Category IV: Seismic Sequence Stratigraphy Interpretation

Sequence stratigraphy is an evolved and refined version of seismic stratigraphy, which reveals depositional mechanism process and architecture of a sequence in minute detail. A depositional sequence is bounded by unconformities and deposited during one cycle of sea level change, where a cycle may be defined by the time interval starting from low sea level, rising to a high, and followed by a fall to low level again. Low and high levels of sea level are defined by its position with respect to shelf of a basin; low if the sea is below the shelf edge and high if it is over it. A geological cycle may vary from millions of years to a few thousands, depending on the order of sea level changes: mega or minuscule. Seismic stratigraphy studies deal generally with third order sequences of cycles of 1–5 million years (my) duration or lower Mulholland 1998. Sequence stratigraphy studies, on the other hand, deal with analysis of small cycles of sea level changes, referred to higher order sequences with cycle duration periods of as little as hundreds of years. The analysis is mostly done from study of outcrops or from an integrated correlation of macro- and micro-scale well log and core data. The duration of deposition being small, the sequences are generally much thinner than lower order seismic stratigraphic sequences and may be difficult to detect on seismic due to limited bandwidth resolution. Nonetheless, help from meso-scale seismic data can be used wherever found relevant. Analysis of stratigraphy of higher order sequences from well log, core and sedimentological data with some help from seismic data may be termed as *seismic sequence stratigraphy*.

Sequence stratigraphy has somewhat different terminologies and nomenclatures compared to those in seismic stratigraphy. The principal emphasis of sequence stratigraphy is on the systems tracts. A depositional sequence defined by seismic stratigraphy is collectively composed of a succession of strata termed as 'systems tracts' deposited during small cycles of sea level change. The systems tracts are mostly divided into three core groups depending on position of sea level with respect to shelf edge and can be individually recognized in well log, core, and sometimes in high resolution seismic data. A very simplified sequence stratigraphy model is described below.

Low Stand Systems Tract (LST)

The sea level is called low stand when it is below shelf edge of a passive margin basin. The bottom part of the sequence deposited over the basal unconformity during low stand is termed a *low stand systems tract (LST)*. The *LST* may be characterized by varied depositional systems which include slope and basin floor fans, fan deltas, submarine channels, prograding wedge complexes and outbuilding deltas depending on the rate of sediment supply from the exposed upland shelf (Neal et al. 1993). *LST*s may be inferred in seismic by external forms of deposits such as mounds, fans, wedges, and by their internal reflection configurations such as chaotic, hummocky and also from prograding clinoforms patterns.

Transgressive Systems Tract (TST)

With the rapid rise of sea level, and with little or no sediment supply envisaged, the shoreline transgresses landward. During this period, the basin may according to one model, accumulate sediments eroded from the earlier deposits of low stand systems tract wedge and deposited backwards as a *transgressive systems tract* (Neal et al. 1993). Ideally, the depositional pattern may be recognized by progressive units of retrostepping seismic sequences bounded by the maximum flooding surface 'mfs', the maximum landward encroachment of the transgressive sea. The 'mfs' demarcates the top of TST from the overlying HST and is often associated with regional deposition of thin veneer of pelagic shale, known

as '*condensed section*'. The '*condensed section*' is considered a good source rock and can be sometimes recognized in seismic as a continuous event, differentiating refection patterns and characters, above and below it.

A sequence may also be described in terms of thinner depositional units known as '*parasequences*'. Parasequences are conformable depositional units bounded by the transgressive sea level known as the marine flooding surfaces *(MFS)*. The units may have patterns of landward back-stepping onlaps, aggradational or basinward progradations depending on sediment supply and changes in sea level. The marine flooding surface indicates sea level rise and separates each of the parasequences. Parasequences are too thin to be perceptible in seismic and are normally interpreted from log and core data. The terms maximum flooding surface (*mfs*) and marine flooding surface (*MFS*) are similar sounding and notation but are used in different context and may not be confused.

High Stand Systems Tract (HST)

As the sea level begins to gradually fall, vast areas of land emerge and large amount of sediment supply is resumed to form *HST* deposits. The resulting parasequences migrate seaward and the high stand deep water deposits can be easily identified in seismic by sigmoid progradational clinoforms, downlapping onto the condensed section. Finally, the sea level falls below the level of deposition and *HST* is exposed to surface erosion and the depositional cycle is completed with the unconformity at the top and a complete sequence is established.

Baum's (1998) growth model of a sequence is shown schematically in (Fig. 32). A representative interpretation of system tracts from seismic based on reflection patterns and external forms is exemplified in Fig. 33. Geologic features like channel and canyon cut and fills, slope/basin floor fans, prograding clinoforms, downlaps etc., can be conveniently inferred from their seismic signatures to interpret linked depositional system tracts. The system tracts comprising the growth

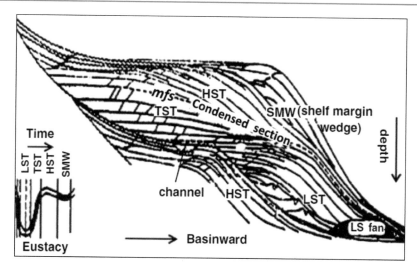

Fig. 32 A schematic growth model illustrating depositional architecture of higher order sequence stratigraphy. The system tracts LST and the HST with intervening TST is represented by '*mfs*' and the '*condensed section*' is marked. Note a cycle is the time span of sea level change between LST-HST-LST which is small and typically a duration of the order of hundreds of years. Seismic stratigraphy studies, on the other hand, deal with lower order sequences of 1-5*my* cycle duration. (Modified after Baum 2)

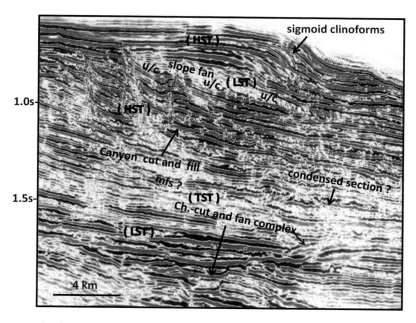

Fig. 33 A representative interpreted example of seismic sequence stratigraphy system tracts based on reflection patterns and external forms of features typically associated with system tracts. LSTs are interpreted from features such as channel cut and fills and slope/basin floor fans whereas HST inferred from canyon cut and fill and sigmoid clinoforms. A continuous downlap reflection between LST and HST is interpreted as 'mfs' and linked condensed section indicating the TST (*Image: courtesy ONGC, India*)

Table 1 Growth model of a Sequence; depositional features and seismic diagnosis

	System tracts	Depositional features	Reflection patterns, external form
Depositional sequence	Unconformity		
	LST: sea level below shelf edge and high sediment supply	Prograding sediments, wedge complex with deltas, slope/basin floor fans, fan deltas, submarine channel cut and fill features, etc.	Shingled/oblique progradational Clinoforms External forms of fans/mounds with chaotic, hummock reflections
	TST: Transgressing sea level and little or no sediment supply	ond Thin retrograde onlaps, topped by a veneer of pelagic shale, the condensed section (*mfs*)	Often are thin sequences, usually below seismic resolution '*mfs*' may be marked at base of HST by the condensed section
	HST: Sea level above shelf edge and high sediment supply	Wedge of progradational stacks with aggradations	Sigmoidal clinoforms downlapping to a weak, continuous reflection (mfs)
	Unconformity		

Note Higher order sets within a sequence, the *'parasequences'* have small depositional durations, are relatively thin and often hard to identify in seismic. Nevertheless, certain depositional characteristics can be traced to typical seismic signatures based on study of reflection patterns and external form of geological features which is tabulated

of higher order sequence, depositional features and their diagnostic patterns in seismic are summed up in Table 1.

The seminal growth model of a sequence described above, however, is based essentially on three important presumptions of depositional environment factors: (1) cyclic deposition of clastics in marine environment, (2) a typical shelf and slope configuration of a passive margin basin, and (3) amount of sediment supply varying suitably with relative sea level changes. Variations in any of the geologic factor concerning tectonics, type of basin, continental depositional regimes (fluvial), rate of sediment supply, and slope of the depositional surface etc., may not form the above described classical growth model of sequence stratigraphy. Such geologic changes are likely to result in drastic modifications in patterns of the depositional sequence and its geometry. Sequence stratigraphy analysis generally appears to work well for passive margin basins with high clastic sediment supply though it is said to be applicable to carbonate deposits also. However, sequence

stratigraphy model depends on geometry of basins and may not be applicable in all types of geologic basins.

Application of Seismic Sequence Stratigraphy

Seismic sequence stratigraphy being a synergistic analysis of well logs, core (biostratigraphic and sedimentological) and high resolution, high density seismic (3D) data, is likely to improve definition of thin reservoirs, especially in subtle stratigraphic traps. The finer details of stratal geometry and related changes in their patterns can also be extremely important for reservoir delineation and characterization, especially with reference to study for presence of small intra-formational paraconformities within the reservoir. *Paraconformity* is a nondepositional unconformity with parallel strata in which the unconformity surface resembles a simple bedding plane. The interfaces may act as horizontal permeable paths but inhibit vertical connectivity

Table 2 Comparison between seismic stratigraphy and stratigraphic interpretation

	Seismic stratigraphy	Stratigraphic interpretation of seismic data
1	Preliminary level of regional interpretation with no/sparse well data; usually made in first phase of exploration	High level synergistic interpretation after calibration with well data; made in later stage of exploration – the delineation and development phase
2	Usually on routinely processed, long regional, wide-grid 2D data	Usually better processed, better resolution, close-grid seismic data (2D/3D)
3	Typically interpretation of entire vertical time section to comprehend basin evolution and evaluation for potential plays and leads	Detailed analysis of prioritized area to upgrade plays and leads to prospects for drilling, limited to prospects and specified time-window of interest
4	For understanding depositional environment and tectonic history	Estimating rock-fluid properties as inputs for estimate of hydrocarbon volumes
5	Regional mapping of structures	Detailed mapping of prospects/reservoirs
6	A powerful tool for evaluating basin for hydrocarbon potential and prognosticate resources	For prospect generation and appraisal for hydrocarbon; reservoir delineation and characterization input to reservoir modeling

Note Though the two expressions, i.e. Seismic stratigraphy and Stratigraphic interpretation, sound alike they are not synonymous. The techniques differ greatly from each other in several aspects, which are highlighted here

within a reservoir sequence, which can influence flow of fluid during production and/ or water injection (Chapter "Evaluation of High-resolution 3D and 4D Seismic Data").

Seismic Stratigraphy and Stratigraphic Interpretation

Seismic stratigraphy and stratigraphic interpretation are two terms widely used, which sound similar, but may not be synonymous. There are, in author's opinion, logical differences between the two approaches; nevertheless they are complementary to each other in achieving the common goal of seismic data interpretation and evaluation. The differences in their core objectives are summarized in Table 2.

References

Balch AH, Lee MW, Miller JJ, Ryder RT (1981) Seismic amplitude anomalies associated with thick First Leo sandstone lenses, eastern powder River Basin. Wyoming, Geophysics 46:1519–1527

Baum RG, Vail PR (1998) A new foundation for stratigraphy. Geotimes 43:31–35

Etris EL, Crabtree NJ, Dewar J (2002) True depth conversion: more than a pretty picture, CSEG Recorder, v 26, pp 1–19

Galloway WE, Yancey MS, Whipple AP (1977) Seismic stratigraphic modeling of depositional platform margin. Eastern Anadarko Basin, AAPG Memoir 26:439–449

Mitchum RM Jr, Vail PR, Thompson S III (1977a) Seismic stratigraphy and global changes of sea level; Part 2, The depositional sequence as a basic unit for stratigraphic analysis. AAPG Memoir 26:53–62

Mitchum RM Jr (1977) Seismic stratigraphy and global changes of sea level: Part II. Glossary of terms used in seismic stratigraphy, Section 2. Application of Seismic Reflection Configuration to Stratigraphic Interpretation. AAPG Memoir 26:205–212

Mitchum RM Jr, Vail PR, Sangree JB (1977b) Seismic stratigraphy and global changes of sea level; Part 6, Stratigraphic interpretation of seismic reflection patterns in depositional sequences. AAPG Memoir 26:117–133

Mulholland JW (1998) Sequence stratigraphy: basic elements, concepts, and terminology. Lead Edge 17:37–40

Nanda N, Singh R, Chopra S (2008) Seismic artifacts—a case study. CSEG REC 33, View issue

Neal J, Risch D, Vail P (1993) Sequence stratigraphy- a global theory for local success. Oilfield Rev 51–62

Vail PR, Mitchum RM Jr, Thompson S III (1977a) Seismic stratigraphy and global changes of sea level; Part 3. Relative Changes of Sea Level from Coastal Onlap, AAPG Memoir 26:63–81

Vail PR, Todd RG, Sangree JB (1977b) Seismic stratigraphy and global changes of sea level, Part 5, chronostratigraphic significance of seismic reflections. AAPG Memoir 26:99–116

Tectonics and Seismic Interpretation

Abstract

Tectonics deals with natural geologic stress and strain (deformation) affecting the structure of earth's crust and its evolution in geologic times. Tectonics is a major component of depositional environment which controls the depositional style of sediments and their structural morphology, paramount in the gamut of hydrocarbon exploration and production. Assortment of structural deformations associated with different types of stress regimes, show typical manifestations in seismic data that can be expediently studied. A brief description of some common and simple geologic structures resulting from different stress types, extensional, compressional and shear (wrench) and their easy recognition and interpretation in seismic data are described, exemplified by seismic images and graphic illustrations. Study of tectonics, the stress regimes, the resultant deformations and their chronological history can be inferred from seismic and may be termed 'seismo-tectonics', analogous to the term seismic stratigraphy.

Tectonics deals with deformation of the earth crust and its evolution in time in response to natural geologic stress. Study of tectonics helps understand earthquakes, volcanoes, building of mountains and valleys, major erosions and depositions linked to geomorphology. In petroleum exploration, tectonics plays foremost role in hydrocarbon generation, expulsion accumulation and preservation. Analysis of tectonic stress regimes and their stages with chronological history is thus extremely important in understanding the evolution and evaluation of geologic basins for hydrocarbon exploration. Applied to seismic interpretation, knowledge of tectonics helps in issues ranging from elementary seismic horizon correlation to generation and evaluation of prospects to hydrocarbon production, that constitutes the whole gamut of petroleum exploration and production. Seismic reflection correlation, the preliminary but crucial input for preparing maps can be attained more consistently by interpreters familiar with understanding of tectonic styles and reduce correlation subjectivities, particularly across faults done by 'phantoming 'and 'jump correlations'.

Tectonic stress besides affecting physical properties of rocks like elasticity and density also impacts pressure and temperature which influence seismic responses. In orogenic belts of mountainous areas under substantial tectonic stress, the effective pressures can impact the rocks and consequently the seismic properties. Flanks of overthrusts may sometimes indicate higher density and velocity than expected for these rocks present elsewhere under no such horizontal stress. However, in geologic basins, where the horizontal stress, compared to the overburden vertical stress is small, it may marginally affect the rock properties.

Tectonic stress regimes, their chronological history and the resultant deformations in the subsurface features are portrayed by seismic images and can be conveniently analysed. The study of tectonics from seismic data in the perspective frame work of petroleum exploration may be termed *seismo-tectonics*, analogous to seismic stratigraphy, which deals with study of stratigraphy from seismic data. The term is hyphenated to mean different from seismotectonics, which generally refers to study of earthquakes and attendant active faults.

Stress on rocks at any point below earth's surface is a resultant of the geostatic (vertical) stress or overburden pressure, the pore pressure, and the tectonic (horizontal) stresses. Role of vertical stress in causing rock compaction and abnormal pressures and its impact on seismic properties are earlier discussed (Chapter "Seismic Wave and Rock-Fluid Properties"). This chapter mainly deals with horizontal tectonic stress and its effects which can be easily analysed from seismic data. Tectonic stresses, the external geologic forces are related mainly to tectonic plate movements and are categorised as (1) extensional, (2) compressional, (3) shear (or wrench) depending on their major stress orientation. A fourth category, the diapiric (shale and salt diapirs), though strictly belongs to extensional stress regime, is considered here separately because of its typical vertical nature of deformation with unique seismic manifestations.

Each of the stress regimes has an assortment of associated deformations or structures, most of which leave distinct footprints on seismic data permitting clear diagnosis. Analysis of tectonics and related deformations can be made from both seismic sections as well as from the seismic maps. Simple geologic structures that are commonly associated with each stress regime and are easily recognized in seismic data are briefly outlined. Illustrations by way of seismic images are exemplified for easy grasping to help better interpretation of seismic data.

Structures Associated with Extensional Stress Regimes

Compaction/Drape Folds

Compaction folds, also known as drape folds, are formed due to differential compaction of sediments deposited on pre-existing paleo-highs. The structure is recognised in seismic by thinning of time intervals (temporal bed thickness) seen on the crests of structural highs, compared to that on flanks of structures (Fig. 1). The structures can also be confirmed by paleostructural analysis (Chapter "Seismic Interpretation Methods").

Horsts and Grabens

Horsts and grabens are up-lifted low lying blocks occurring between two adjacent normal faults and are well imaged in seismic profiles (Fig. 2). Normal faults are dip-slip faults where the hanging wall (down throw side) has slipped down due to gravity and predominantly occur in extensional stress regime. The structural high/ low blocks between the normal faults, are formed due to relative movement of the fault blocks along the two fault planes. The features usually orient in the direction of regional depositional trend and can be inferred from seismic horizon maps. Paleostructural analysis can also be useful in revealing the paleo trends where the orientation might have changed due to post depositional tectonics.

Normal Faults, Types and Compaction Faults

Normal Faults
Normal gravity faults are caused by deep-seated tectonic stress affecting the basement and the overlying sediments. Normal faults can be categorised as synsedimentary, reactivated or post-depositional (younger) faults (Fig. 3). The age, history and type of the faults can be easily studied in seismic segments from patterns and orders of displacement of faults based on

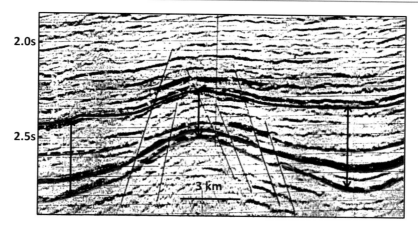

Fig. 1 Seismic example of drape(compaction)folds associated with extensional stress. Drape folds are caused by compaction of sediments deposited on paleo highs and are Identified in seismic by lesser bed thickness at the crest compared to the flanks (shown by arrows). Note the gradual flattening of structural relief upwards at shallower depths due to lesser effect of compaction at shallow depths (Image: courtesy ONGC, India)

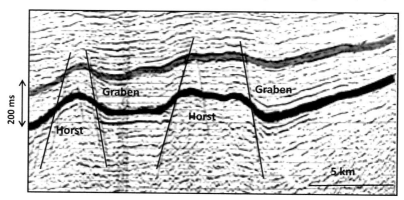

Fig. 2 Showing seismic expression of horst & graben structures in extensional stress regime. The horst is the up-thrown and the graben down-thrown block caused by two adjacent gravity faults as a result of relative block movement between the faults (courtesy, ONGC, India)

variations in sedimentary thickness on either side of faults, the foot wall and the hanging wall.

Synsedimentary faults are faults concurrently active with sediment deposits and are recognised by increased sedimentary thickness on the hanging wall compared to the foot wall (Fig. 3a). The difference in temporal thickness on either side can in fact, indicate broadly the amount of subsidence the hanging wall has gone through.

Reactivated faults are earlier existing faults reactivated (rejuvenated) at later time and can be identified by differential displacements (throw) of faults, being more for the older beds compared to younger beds. This is due to added displacement

to that of existing fault during reactivation (Fig. 3b).

Post-depositional faults, in contrast to synsedimentary faults, show equal fault throws (displacements) at all levels. This signifies the faults occur after deposition and are also sometimes referred as young faults, the age of the fault decided by the youngest bed affected (Fig. 3c).

Compaction Faults

Compaction faults are typically low-angle rootless faults, which do not extend deeper into older sections (Fig. 4). The faults mostly occur in thick formations of finer clastics during their deposition due to large amount of compaction taking place

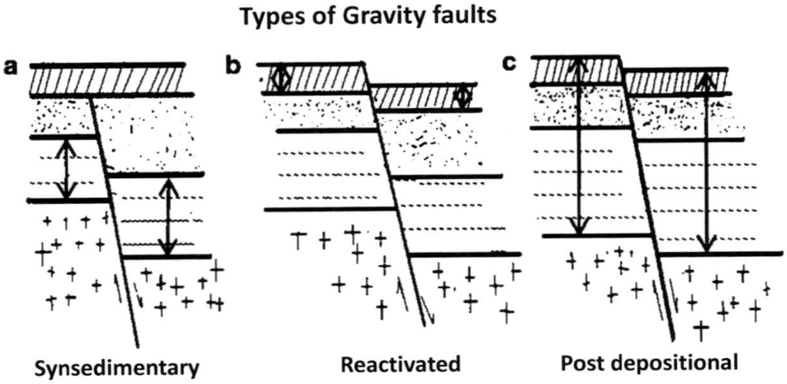

Types of Gravity faults

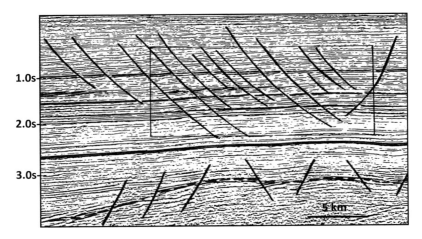

a   **b**   **c**

Synsedimentary   **Reactivated**   **Post depositional**

Fig. 3 Schematic illustrating types of normal faults (**a**) Synsedimentary fault shows greater sedimentary thickness on hanging wall compared to foot wall indicating it was active concurrently with deposition (**b**) Reactivated fault is an earlier fault reactivated after deposition during its burial. As a corollary, faults which do not show such decrease in dip with depth may be inferred as occurring post-compaction.

of younger beds (hatched) resulting in increasingly larger fault throws for the older beds (existing plus recent) than the younger ones. (**c**) Post deposition (young) fault is typified by similar throw (displacement) at all beds and its age decided as post the youngest affected bed

Listric and Growth Faults and Roll-Over Structures

Listric faults are rootless normal faults, discernible on seismic sections by their concave upward geometry and by a gradual progressive decrease of dip with depth, until they merge with the bedding plane. Synsedimentary listric faults, characterised by thicker sedimentary sections (growth) observed on the hanging wall are known as *growth faults*. A schematic expression of a typical growth fault is shown in Fig. 5. It may be stressed that all listric faults are not growth faults as are termed loosely sometimes. Substantial depositional load on the subsiding

1.0s

2.0s

3.0s

5 km

Fig. 4 Seismic segment displaying compaction faults. Compaction faults are post-depositional faults associated with extensional stress. The faults mostly occur in thick clastic sections with dips of fault plane gradually decreasing downwards. They do not extend to deeper depths and are called rootless faults. Note the two different non-related fault systems, compaction in shallow and normal gravity faults in deep horizons with a strong horizon in between without affected by faults (Image: courtesy ONGC, India)

hanging wall of a growth fault leads to instability, ensuing sliding of sediments along the fault plane leading to a fan like structure imaged in seismic (Fig. 6). Under certain geologic conditions, the sliding sediments (the 'fan') may form roll-over structures against the fault plane which can be seen in seismic (Fig. 7). The 'fans', however, are sometimes referred as 'roll over' structures though they do not show the reversal of dips and can be misleading as a potential structural trap.

All growth faults, however, do not end up in roll-over structures. It depends on other factors such as sedimentary load, mobility of the clastic sediments and the geometry of the growth fault plane. It may also be noted that growth faults dip basinward and the sediment supply from land side is loaded on the hanging wall as it allows accommodation space due to continuing subsidence caused by gravitational sliding of sediments along the fault plane.

Cylindrical Faults and Associated Toe Thrusts, Imbricates

Cylindrical faults appear commonly at basin slope margins and are caused by excessive loading of finer clastics with high mobility on the slope (Fig. 8). This leads to slope failure with attendant gravitational slides of the mobile sediments along the fault plane, also known as 'gravity tectonics. Often a common feature associated with this kind of fault is a thrust (described later) at the distal end of the fault. These faults are known as *'toe thrusts'* which are caused by localized compressional stress created by resistance within the mobile shale mass to move further. In extreme cases, intense gravity tectonics may also cause huge slides that can create *imbricates* (described later on), that are clearly imaged in seismic dip sections (Fig. 9). It is important to differentiate these structures from thrusts and imbricates that are the typical deformations associated with compressional stress.

Structures Associated with Compressional Stress Regimes

Folds

Compressional folds form due to shortening of formation under compressional stress. In seismic profile the folds typically show equal or larger time-intervals (thickness) at the crest of the structure compared to the flanks (Fig. 10). This is

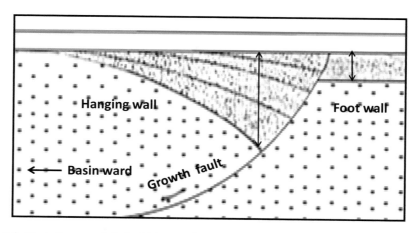

Fig. 5 Schematic illustrating growth fault which occur in extensional regime. Growth fault is rootless synsedimentary listric fault with upward concave geometry and the fault plane dip progressively decreasing with depth till it merges with bedding plane. The concave geometry faces basin ward and the sediments continue to pile (growth) on the hanging wall block as it creates accommodation space due to ongoing gravitational sliding

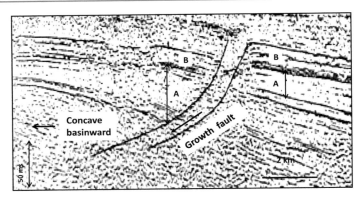

Fig. 6 Seismic segment example of a growth fault, characterized by Increased thickness (growth) of bed 'A' on the hanging-wall with the concave plane facing basin ward. In contrast, note the near-similar thickness for bed 'B' on either side of fault suggesting reactivation of the fault after deposition of 'B'

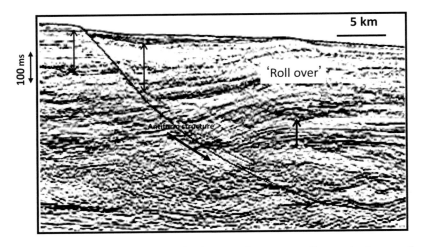

Fig. 7 Seismic image displaying growth fault related 'roll over' structure in extensional stress regime. Over-loading On the hanging wall block leads to instability and sliding of sediments along the fault plane which sometimes cause roll-over structure, contingent to other factors such as the fault plane dip, sediment load and mobility of the clastic formation. (Image: courtesy ONGC, India)

in stark contrast to expression of compaction folds evinced in seismic.

Reverse Faults, Thrusts & Overthrusts

Reverse faults, thrusts and overthrusts are typical deformations caused by compressional stress. As the name suggests, the reserve faults are opposite to the nature of normal faults, the hanging wall moves up relative to the foot wall. Low angle reverse faults are referred as 'thrusts' and very low-angle thrusts as 'overthrusts'. Deformations due to compression may or may not involve the basement. Where the basement is not involved, the thrust planes are restricted to the sedimentary section. The fault plane dip decreases with depth till they merge with underlying bedding plane similar to a normal listric fault. Such deformations are termed *'thin-skinned* tectonics and the plane where the thrust merges is known as *'decollement'* (Fig. 11). Decollement planes are usually within compliant strata that tends to absorb the stress.

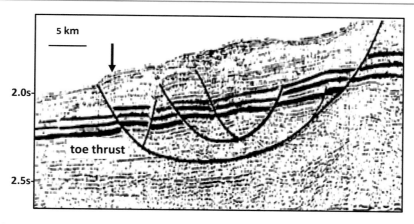

Fig. 8 Seismic image showing a cylindrical fault which occurs in extensional stress regime at slope margins in deep waters. The fault is caused by excessive loading of finer clastics resulting in slope failure and attendant gravitational sliding of the high mobility-sediments along the fault plane. The cylindrical fault is often typified by small thrusts (shown by arrow)at the distal end known as 'toe thrust'. Toe thrusts are caused by internal resistance offered by the mobile shale for further movement. Note, the thrust is caused by localised compression within the shale mass and not due to regional compressional stress. (Image: courtesy ONGC, India)

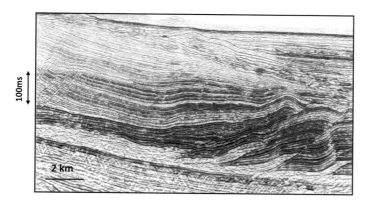

Fig. 9 Example of seismic image showing imbricate structures formed in extension regime. In extreme cases of gravity sliding of huge mass of mobile shale due to slope failure, buttressing force (compression) within the shale deforms to imbricate structures. It is also called "gravity tectonics" and must be distinguished from imbricates formed in compressional regimes. (Image: courtesy ONGC, India)

An important aspect of reverse faults and thrusts is understanding the phenomenon of movement of block along the fault, known as *'Vergence'*. The determining factors for *vergence* of a fault under extensional and compressional stresses include paleo topography, pre-existing faults and weak zones in the rocks under stress and is illustrated graphically (Fig. 12). Formations where faults preexist, under extensional stress, the vergence is decided by the hanging wall going down under gravity (Fig. 12b), whereas under compression the hanging wall moves upward (Fig. 12c).In near flat formations with no pre-existing faults it may be difficult to determine the vergence where the topography or the weak zones in the rocks may influence the vergence (Fig. 12a, d, e and f). *Vergence* can be easily inferred from the seismic sections that can help characterize the tectonic history to recreate the paleo structural configuration, an important factor in evaluation of hydrocarbon prospect. Understanding vergence also can explain apparently

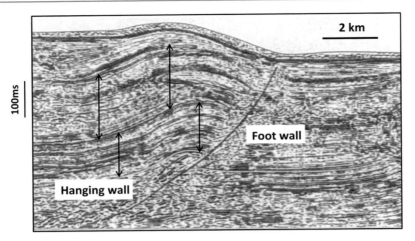

Fig. 10 Seismic image displaying expression of a fold and reverse fault associated with compression. The fold is typically identified by larger or similar bed (temporal) thickness at crest than at flanks (shown by arrows). Note the contrasting characteristics of bed thicknesss to drape folds in extensional stress regime. (Image: courtesy, ONGC, India

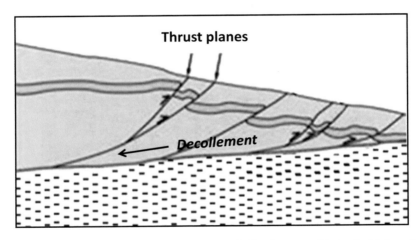

Fig. 11 A schematic diagram illustrating 'skin tectonics' in compressional regime which does not involve basement. The dips of the thrust planes (marked by arrows) gradually decrease with depth till they disappear in an incompetent basal plane, called the '*decollement*' without extending to depths

baffling issue where seismic shows a normal fault at deeper level but as reverse fault at shallower levels (Fig. 13). This happens because the hanging wall of the pre-existing normal gravity fault, under compression, gets reactivated and moves up resulting in reverse faults at younger levels but its upward displacement is not enough to over compensate the throw at the deeper level. It thus remains as a normal fault though with reduced displacement than it had initially.

Structures Associated with Shear Stress (Wrench) Regimes

Wrench tectonics is caused by horizontal shear stress couples due to differential movement of deep crustal blocks and thus necessarily involves the basement. Stress couples can be divergent (transtensional) or convergent (transpressional) depending on the dominant stress component of

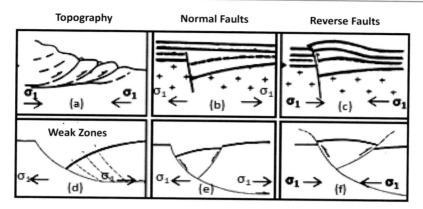

Fig. 12 A schematic illustrating phenomenon of 'vergence' of a fault which is movement of block under tectonic stress. In compression, it is the hanging wall of a fault which always gets pushed up. With no preexisting faults or flat beds, the movement is controlled by topography (**a**) or existing faults/weak zones in the rock (**d**).The arrows and symbol σ_1 denote the maximum principal horizontal stress axis. The vergence is shown for compressional (**a**, **c** and **f**) and extensional stress (**b**, **d** and **e**) where the hanging wall moves downwards due to gravity

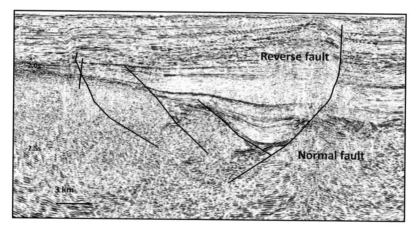

Fig. 13 Example of seismic image displaying a fault which is normal at depth but reverse at shallower levels. Reactivation of the normal fault during the compressional regime caused the hanging wall move upwards but not enough to negate fully the preexisting displacement of the normal fault at depth and therefore remains as a normal fault though with reduced throw than the initial throw (Image:courtesy ONGC, India)

extension or compression. The resulting structures may accordingly vary between the two extreme cases of wrench, depending on the dominance of extensional and compressional stresses, known as transtensional and transpressional stress.

En Echelon Conical Folds

Conical folds are compressional folds created at an inclined angle to stress orientation with *en echelon* pattern. An *en echelon pattern* is a configuration of number of parallel to sub-parallel features occurring with progressive lateral offsets, the term mostly referred to folds and faults. *En echelon* conical folds are a series of folds, laterally offset, and are typical deformations associated with wrench tectonics. The patterns of fold axis orientation and their lateral offsets indicate the shear stress direction. It is called sinistral or left lateral shear (anticlockwise) when the fold axis tilts and off- shifts are progressively to right (Fig. 14a) and is called dextral or right lateral (clockwise), when the axis and the shift are to the left (Fig. 14b).The folds are often associated with high angle vertical faults that can be clearly seen in seismic sections (Fig. 15). However, it may be difficult to identify conical folds on seismic

profiles but can be seen clearly in planview, mapped as a number of discrete fold culminations shifted laterally in *en echelon* pattern.

Half Grabens with High-Angle Faults

High-angle, near vertical faults and associated half grabens are formed in divergent (transtensional) wrench regimes. A half graben is a sag feature that forms against a major fault present only on one side, unlike two faults on either side of a graben (Fig. 16). Half grabens associated with low-angle faults, however, are formed commonly in rift basins in extensional stress regime.

Sinistral (Anticlockwise)   **Dextral (Clockwise)**

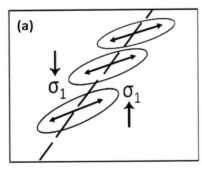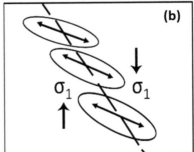

Fig. 14 A plan view schematic illustrating wrench associated enechelon conical folds, progressively with lateral offsets. The inclination of the fold axis tilt with respect to principal stress direction and its lateral offset pattern indicate the shear stress direction. (**a**) Inclination of fold axis and progressive lateral shifts to right signify left lateral (sinistral) wrench while (**b**) fold axis inclination and lateral shifts to left indicate right lateral wrench (dextral). The principal stress directions are indicated by arrows

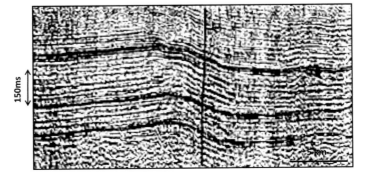

Fig. 15 Seismic expression of a compressional fold structure with high angle vertical fault, associated with divergent (transtensional) wrench. Note the similar bed thickness at crest and flank of structure. (Image: courtesy ONGC, India)

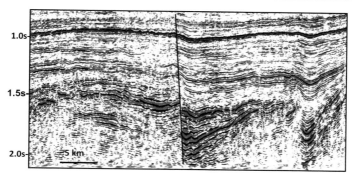

Fig. 16. Displaying seismic expression of high angle near-vertical fault forming half graben associated with transtensional wrench. (Image: courtesy ONGC, India)

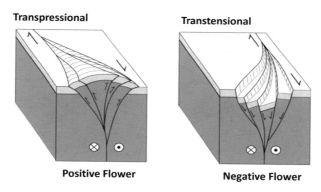

Fig. 17 Schematic diagram illustrating 'flower structures' associated with wrench. (**a**) Positive flowers are antiforms created with reverse faults signifying transpression (convergent wrench) and (**b**) Negative flowers are synforms with high-angle normal faults indicating transtension (divergent wrench)

Flower Structures and Inversions

Flower structures are typically associated with wrench tectonics and are clearly visible in seismic sections. Antiforms associated with reverse faults are known as positive flower (Fig. 17a) and synforms with normal faults as negative flowers (Fig. 17b) depending on type of dominant wrench component, convergent (transpressional) or divergent stress (transtensional). Seismic expression of a negative flower structure sometimes is referred as 'inversion structure' and an example is shown in Fig. 18.

Strike-Slip Faults

Strike-slip faults are faults characterized by considerable horizontal displacements compared to vertical, which often make them difficult to identify on normal vertical view seismic sections. Nevertheless, indications of strike-slip faults can be recognised in planview where the map may show en *echelon* faults, called *'dog-leg'* pattern. The subtle transverse faults joining the major normal faults in planview are the displacement of strike slip faults and can be mapped by horizontal slice attributes from 3D seismic high resolution

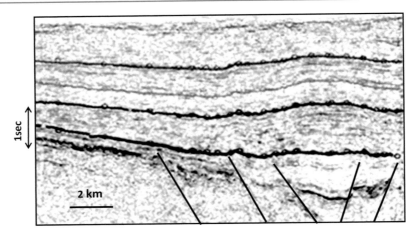

Fig. 18 Seismic example of negative flower with antiforms overlying a synform. Note the larger bed thickness of the fold at shallow level, overlying synform which is affected by high angle normal faults. This typically suggests negative flower structure under transtensional stress. The feature is also referred sometimes as 'inversion structure'. (Image: courtesy ONGC, India)

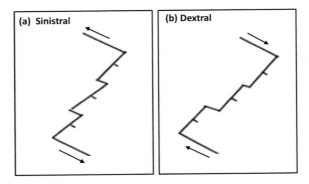

Fig. 19 A sketch illustrating 'dog leg' fault pattern in plan view, the enechelon normal faults progressively off-shifted suggesting strike slip faults. The lateral shift pattern indicates shear direction. (**a**) Progressive lateral shifts to left indicates anticlockwise (sinistral) and (**b**) right shifts as clockwise (dextral) stress

data (Chapter "Analyzing Seismic Attributes"). The *dog-leg* fault pattern also offers clues to the direction of the shear stress, dextral (Fig. 19a) or sinistral (Fig. 19b) depending on the direction of off-shifts of the faults, towards left or right.

Further evidence of wrench tectonics can be found in seismic maps by anomalous trends of structural features and contours. Convergence of a high trend into low and vice-versa, anomalous juxtaposition of counterpart of a structural high against a low or the other way round, across a fault, and sudden change in trends of contours are some of the distinctive evidence of wrench

stress (Fig. 20). Abrupt variation in data quality in the same or adjoining seismic profiles showing uncorrelatable seismic events and large scale mismatches can also be a strong clue to existence of strike-slip faults.

Salt/Shale Structures

Shale and salt deformations classically belong to group of structures under extensional stress regime but are categorized separately because of their typical manifestation in seismic, the

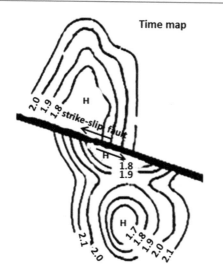

Fig. 20 Example showing typical indications of strike-slip faults in seismic map. Note the counterpart of the 'high' at top of the map is juxtaposed across the fault by a laterally shifted high with completely different orientation, characteristic of strike slip fault. The relative movement (slip) of the two high features along the fault (shown by arrows) signifies the stress direction as, sinistral. (Modified after Moody 1973)

vertically piercing features within sediments. Widely spread thick layers of salts in a basin undergo plastic flow due to overburden stress under sedimentation and assisted by density differences with the surrounding sediments forms upward growing structures known as 'diapirs'. Mobile compacted shales also exhibit similar diapiric structures which are termed variously as 'shale pillows', 'mud diapirs', 'mud volcanoes' and 'mud flows'. Shale and salt tectonics are similar, though shale diapirs are characteristically less intensive and nonintrusive in nature. Nonetheless, shale and salt structures can be discriminated on the basis of seismic velocity, salt having a much higher velocity relative to the surrounding sediments than that of shale. The shale and salt diapirs are also linked to distinctly different depositional environment and can be interpreted on the basis of geological history of the basin. Mobile uncompact shales usually are deposited in the distal part of the basin, beyond the shelf edge, and deform into assortment of structures commonly occurring in deep waters (Fig. 21). However, while salt diapirs and related structures are often of exploration interest, the shale diapirs, in contrast, are mostly avoided. The shale structures are also usually associated with overpressure and are occasionally gas charged and are considered potential drilling hazards.

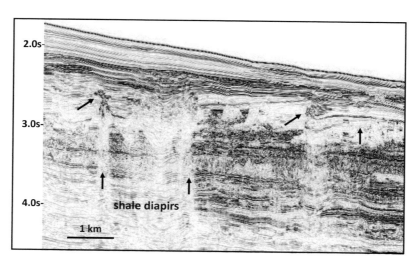

Fig. 21 A seismic image showing typical shale diapirs and assortment of other shale structures (indicated by arrows) in offshore super deep water. Note the diapirs vertically piercing through the sediments suggesting their occurrence as of late age before deposition of the strong, continuous young reflector, overlying it. (Image: courtesy, ONGC, India)

Salt Diapirs—Types

The study of salt tectonics, which includes the process of salt flow and the resultant deformations that create salt structures, is known as *'halokinesis'*. Salt diapirs are growths, which may build up to kilometres in height. The diapirs can be categorised as synsedimentary or post-depositional, similar to terming of faults, depending on history of their growth. Growth of diapirs episodically concomitant with process of sedimentation is known as synsedimentary diapirs and growth occurring at later time after deposition of sediments, is known as post-depositional. Further, the diapirs can be intrusive or extrusive depending on the growth remaining below the depositional level or getting exposed. Seismic images of salt diapirs provide excellent clues to unravel the chronological history of their development. A schematic figure illustrating the growth process with time line is shown in Fig. 22. Intrusive synsedimentary diapirs show draping of overlying younger beds formed during the interim period of growth episodes (Fig. 22a). Whereas extrusive growths, in contrast being exposed at some point of time,

would show erosional unconformity and deposition of near-flat younger strata against flank (Fig. 22b). Post depositional diapirs, on the other hand would show vertical piercement into the overlying beds resulting in considerable drag of these beds at diapir flank (Fig. 23). Further, the age and intensity of the period of diapir growth and the time when it ceased can also be interpreted from the seismic image based on the degree of bed-drags and appearance of drape folds on top. Intrusive growth of diapirs are often associated with attendant radial faults at the top of the diapir known as 'adjustment faults' (Fig. 24). Adjustment faults occur as a result of release of accumulated stress stored during the growth. In case of extrusive diapirs, no such faults may be evinced as the stress gets released due to erosion during its exposure at the surface.

Salt diapirs sometimes form attendant structures known as 'flank *synclines*' and '*turtle back*' structures. The deformations are caused due to withdrawal of salt from the flanks that feed the continuing vertical growth of salt. Removal of salt from the basal part causes the overlying beds to collapse, resulting in forming flank synclines (Fig. 25). Continuous feeding of salt from the

Fig. 22 A schematic illustrating syn-sedimentary growth of salt mass with continuing deposition of overburden sediments. (**a**) The intrusive synsedimentary diapir is typified by drapes in overlying beds and (**b**) the extrusive diapir growth by presence of erosional unconformity with the near-flat younger strata terminating against the vertical diapir

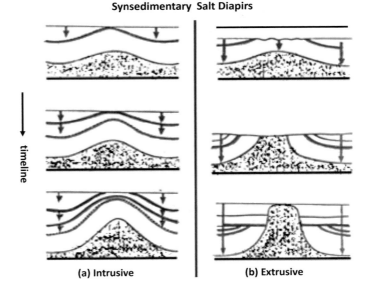

Synsedimentary Salt Diapirs

timeline

(a) Intrusive   (b) Extrusive

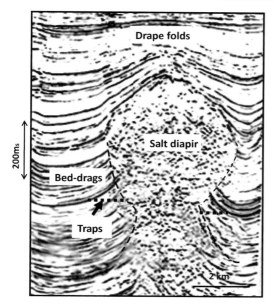

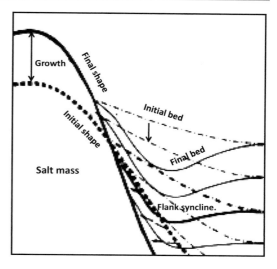

Fig. 23 Showing seismic image of a post depositional salt diapir triggered by overburden loading. Note the upward drag of beds due to the piercing salt diapir and the resultant formation of stratigraphic traps (marked with dotted line) against the flanks with salt behaving as lateral seal. The boundary of salt flank as interpreted(dotted lines) is clearly subjective and dubious. The uncertainty in imaging the salt dome geometry enhances greatly the risk of drilling the prospects

Fig. 25 A schematic diagram illustrating 'flank syncline' structures due to continuing growth of salt. The growth (arrow) is fed by withdrawal of salt from flanks triggered by overburden pressure and removal of salt from the base leads the overlying beds to collapse which results in forming synclines at flanks of a salt diapir

flanks which builds massive vertical growth of diapir, besides forming 'flank synclines' on both sides, sometimes form 'humps' in between

space, resembling back of turtle and called 'turtle back' structures (Fig. 26).

Salt diapirs commonly attract explorationists because of their associated structural and strati-structural traps. Drape folds in the overlying sediments and strati-structural traps formed at the flanks of diapirs with the salt-core acting as lateral seal can be prospective traps for hydrocarbon

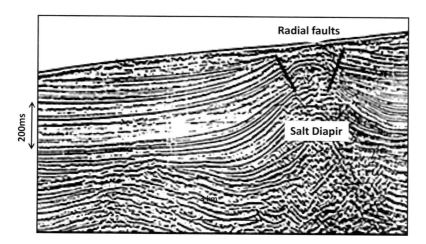

Fig. 24 A seismic example of a synsedimentary intrusive salt diapir, concurrently growing without getting exposed to surface. The continuing build-up stress within the salt

diapir gets released at some point of time through creation of faults at the top of structure. Note the typical radial faults at the top of the diapir

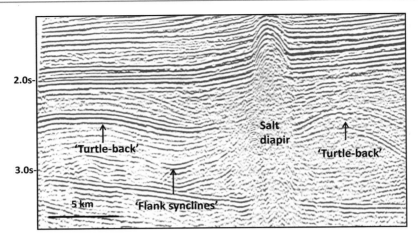

Fig. 26 A seismic example showing vertical piercing of salt diapirs and ensuing structural deformations, the 'flank synclines' and the associated 'turtle-back' structures. Feeding from adjacent flank salt diapirs (not seen here), causes flank synclines due to removal of salt which may result in forming hump features resembling 'turtle back' and are called 'turtle-back structures (After Anstey 1977)

accumulation. However, the diapir flanks are needed to be precisely defined as geometry of the strati-traps, abutting updip against the salt flank, depends on the slope and disposition of the salt body (Fig. 23). Though diapirs can be strikingly well captured in seismic images, precise delineation of the salt flank geometry may be problematic. Referring to Fig. 23, definition of the salt diapir from the seismic is highly dubious which can be critical in deciding the exact location for drilling wells for hydrocarbon. The small lateral extent of the traps and the steep dip of beds allow very little margin of error for drilling at the proper location; structurally too high may miss the reservoir and too low may hit water.

Significance of Seismo-tectonics in Seismic Data Evaluation

Analysing tectonic stress regimes, their style and chronological history from seismic provides important insights into the genesis, timing and morphology of tectonic structures which helps proper evaluation of hydrocarbon prospectivity in the area. Geologic consistent correlation of seismic horizons is the crucial input for preparing accurate maps, the key to hydrocarbon exploration, based on which prospects are generated, evaluated and reserves estimated for their hydrocarbon prospectivity. From deciding the optimal drilling locations to estimating hydrocarbon reserve volumes, through all stages, exploration ventures depend on the accuracy of the seismic maps and this cannot be fully attained without good working knowledge of tectonics. Types of faults, their hades and throws are useful in correlation of seismic events across faults and in estimating proper thickness of sediments on both, the foot wall and the hanging wall. Vergence, another important element of fault can be made from seismic to assist in reconstruction and balancing of sedimentary sections in complicated geologic regions, necessary to comprehend proper evolution and evaluation of basins in general and prospects in particular for exploration.

Assimilation of tectonic and depositional styles and their architecture helps form a tectono-stratigraphic framework of a basin, the bedrock on which petroleum system modelling is built, an important tool used in petroleum prospecting and discussed in the next chapter. Knowing the tectonics can also be geologically advantageous for comparing with similar known petroliferous basins elsewhere in the world to help evaluate hydrocarbon potential by drawing analogies. For instance, recognising wrench tectonics foot prints in seismic in an area, can by itself, be a precursor to certain favourable aspects of petroleum geology, inherently associated with wrench tectonics. Consider a case where the basement fault systems are active for a long period of time in a wrench regime that would facilitate differential block movements of varying degree with associated subsidence, tilting and rotation of blocks. This could create situations conducive for creating possibility for a myriad variety of potential petroleum habitats through source–reservoir connectivity either by lateral juxtaposition or vertically through faults. Furthermore, wrench related deformations are more likely to offer an assortment of interesting structural and stratigraphic plays that can be suitable for hydrocarbon accumulation, enhancing prospectivity potential of a basin.

Comprehending the dynamics of tectonics is critical for prospect appraisal to properly evaluate the prospective traps such as the anticlines, fault closures, wedges and unconformities for their effectiveness to trap and preserve hydrocarbons. Faults, specially, are extremely important in this regard and warrant careful detailed

synthesis as they are known to play major roles in migration and accumulation of oil and gas. Faults, attributes and their roles are in particular, important in petroleum exploration and can be analysed from seismo-tectonic studies and are discussed in detail in the next chapter.

References

Anstey AN (1977) Seismic interpretation, the physical aspects, record of short course "The new seismic interpreter", IHRDC., 5–27 to 5–29 .

Billings MP (1972) Structural geology, 3rd edn. Prentice-Hall, Englewood Cliffs, New Jersey

Badgley PC (1965) Structure and tectonics. Harper and Row, New York, pp 1–521

Crans G, Mandl G, Harrembourre J (1980) On the theory of growth faulting: a geomechanical delta model based on gravity sliding. J Pet Geol 2:265–307

Harding TP, Tuminas AC (1989) Structural interpretation of hydrocarbon traps sealed by basement normal faults at stable flanks of foredeep basins. AAPG Bull 73:812–840

Moody JD (1973) Petroleum exploration aspects of wrench-fault tectonics. AAPG Bull 57:449–476

Wilcox RE, Harding TP, Seely DR (1973) Basic wrench tectonics. AAPG Bull 57:74–96

Seismic Stratigraphy and Seismo-Tectonics in Petroleum Exploration

Abstract

Integrated studies of seismic stratigraphy and seismo-tectonics provide a highly effective and indispensable tool used in petroleum exploration. The unified approach helps comprehend basin evolution & evaluation, organises 'tectono-stratigraphic' framework for building basin petroleum system modelling (BPSM) and generate and evaluate potential plays and prospects for exploratory drilling.

Petroleum system modelling deals with assessing potential of each component required for hydrocarbon discovery, i.e. the source, reservoir, migration, trapping, accumulation and preservation and helps estimate hydrocarbon resource potential for virgin / underexplored basins. The modelling study provides better insight to the hydrocarbon accumulation system and often is used to find new and more hydrocarbon with less risk in matured petroliferous basins. It also incites new geologic play concepts leading to upgradation of prospects and prognosticated resources to re-strategize basin exploration policies.

Utility of unified analysis of seismic stratigraphy and seismo-tectonics for basin evolution and hydrocarbon evaluation along with petroleum system modelling is described. The geologic, seismic and geochemical aspects of assessing hydrocarbon source quality and thermal maturity for generation is introduced.

Prospect generation and appraisal, estimate of volumetric reserves and techno-economic analysis are highlighted. Since faults play major roles as conduits, seals, and leaks in hydrocarbon migration and accumulation, fault attributes analysis and fault seal integrity studies for traps are also included in the chapter.

Seismic stratigraphy and seismo-tectonics are highly effective interpretation techniques, indispensable in the quest for discovering hydrocarbons. Integrated, the unified technique promotes understanding the evolution history of geologic basins. The analysis helps in reliable evaluation of geologic plays, leads and prospects for hydrocarbon exploration. Plays are terms used to indicate possibilities of hydrocarbon occurrence by qualitatively assessing the major components for petroleum accumulation, i.e. the source, reservoir and trap in a basinal scale. The leads are prospects that are identified but not fully assessed, whereas prospects are the traps which are properly evaluated including techno-economics for undertaking drilling. Though, the terms may appear as semantics, it is important to be aware of the terms, used contextually in petroleum industry.

The technique of combined analysis of seismic stratigraphy and seismo-tectonics is particularly useful in virgin or least-explored areas, where well data do not exist or are scrappy and

N. C. Nanda, *Seismic Data Interpretation and Evaluation for Hydrocarbon Exploration and Production*, Advances in Oil and Gas Exploration & Production, https://doi.org/10.1007/978-3-030-75301-6_5

inadequate. The degree of success in predicting hydrocarbon discovery, however, depends on skill and experience of the interpreter in comprehensive understanding of fundamentals of tectonic styles and depositional systems. Individual as well as complementary roles played by the two techniques in prospect evaluation and building tectono-stratigraphic framework for petroleum systems modelling is stressed.

Basin Evaluation

Seismic data interpretation by synergising seismic stratigraphy and seismo-tectonics, reveals the depositional and tectonic history of sedimentary fill of a basin. This helps to recognize and map potential geologic plays, leads and prospects for assessing hydrocarbon prospectivity. Basin evaluation is carried out by assessing the elements of petroleum systems—a term used for the geologic elements and processes responsible for hydrocarbon accumulation and is often achieved by modelling, commonly known as petroleum system modelling and is discussed later in the chapter. The crucial components that define petroleum system are (i) the source and generation; (ii) reservoir; (iii) migration (pathways and timing) and (iv) entrapment and preservation. Each element is briefly examined and application of seismic stratigraphy and seismo-tectonics tools to evaluate them is stressed.

Source and Generation Potential

The hydrocarbon potential of a basin is assessed based on knowledge of source and its generation capacity. The source potential is contingent to the amount of organic-rich fine clastics (shale), total organic content (TOC) and the kerogen type. Hydrocarbon generation capability depends on the degree of thermal maturity that is controlled by the depth, temperature, pressure (burial history) and period of time duration. Kerogen is an organic matter, which under pressure and temperature is transformed to oil or gas, depending on its type. More about kerogen and thermal maturity is discussed later in the chapter.

Source rocks are deposited in a variety of environments such as fluvial, deltaic, lacustrine, shallow marine and deep-marine and with varying characteristics. Lacustrine, deltaic and shallow marine shales are considered better source potentials, due to likelihood of more TOC content and the type of kerogen to produce oil than the fluvial or deep marine shales. Seismic facies analysis identifies the depositional environments, their associated shale facies, extent, thickness and distribution. This helps assess the amount (volume) of source rock, the organic matter available with likely TOC and kerogen type to evaluate the quality of source for generation of oil and gas. Potential source rocks also require preservation of organic matter and consequently shales deposited under anoxic environment are preferred. Whereas, source deposits in oxic environment are considered poor due to oxidation of carbon matters, exposed on surface without undergoing subsidence. Oxic and anoxic environments can also be predicted by seismic stratigraphy and seismo-tectonic analysis. Unconformities, erosional surfaces and rate of subsidence of source rocks which influence oxic and anoxic environments can be inferred from seismic evidence of basinal tilts, subsidence and faults types with their genetical history.

The generation potential of a source is determined by assessing thermal maturity, a process which transforms kerogen to hydrocarbon. Thermal maturity elements are generally linked to depth of burial which offers clue to required pressure, temperature and the duration period. Seismo-tectonics and seismic facies analysis unravel the clues to burial history by analysing genesis, age and history of faults. For instance, consider the implications of a synsedimentary fault and active episodically for a long period of time, analysed from seismo-tectonics studies. Identifying the fault as synsedimentary not only signifies the increasing thickness (volume) of source rock (shale) on the basin side but also on its deposition under anaerobic (anoxic) condition. It also indicates the other important element, the burial history. Seismic studies can broadly reveal the individual tectonic episodes, recurring subsidence and its rate, time lapses between the episodes and finally the depth attained during a geologic period. These are factors generally considered for potential source generation and maturation. The depth also signifies the temperature though it depends on ambient thermal gradient in the area. Thermal gradients vary in areas and are likely to be more intense in tectonic zones. Inferences of severely tectonized zones and intrusive volcanic activities from seismic analysis can also offer inklings to geothermal gradient of the area. However, it may be stressed that more than generation, the quantity of hydrocarbon expulsed, is important which ultimately decides the amount of hydrocarbon accumulated in the traps.

Reservoir Facies

Reservoir rocks with good intergranular (primary) porosity and permeability are associated with high energy fluvial, deltaic and shallow marine environments, mostly deposited on basin shelf and slopes. Reefal mounds on shelf edges also show excellent primary porosities. Primary porosities are the initial voids when the rock is deposited, whereas secondary porosities are created subsequent to deposits (Chapter "Seismic Wave and Rock-Fluid Properties"). Secondary porosities are more common in carbonate rocks and are often assisted by unconformities, erosion and faults. Carbonate structures, exposed to weathering can show good secondary porosities and behave as potential reservoirs. Seismic sequence recognition of unconformities and seismic facies analysis of external forms and internal reflection configurations can identify carbonate mounds, reefs and other high energy clastic reservoir facies such as delta lobes, fans, channel cut and fills. Seismo-tectonic studies indicate uplifts, faults and their chronological history and provide clues to unconformities that

can potentially induce secondary porosities such as leaching, channelling, vugs and fractures and augment permeability for production. A unified analysis thus can be immensely useful in predicting depositional and tectonic elements for evaluating reservoir facies, their quality and more importantly their distribution in the basin.

Migration, Pathways and Timing

The hydrocarbon migration is a two-step process, dealing with expulsion from the source rock and transmission through carriers to traps for accumulation. Expulsion from source is known as *primary migration* and getting into the reservoir rock is called the *secondary migration*.

Primary Migration

Primary migration is the process of expulsion of hydrocarbon from the source rock. There are several theories about the expulsion mechanism but it is mostly believed to be caused by the high pore fluid pressure created within the generating source itself. As hydrocarbons are generated by thermal cracking of kerogen, the generated fluid tries to occupy larger volume than what was in the original formation. Rise in temperature with subsidence further increases the expansion of fluid volumes resulting in high pore pressures within the confined source. At some point of time the continuing increase in pressure causes micro fractures in the source which facilitate expulsion of hydrocarbon. Episodes of subsidence and high-pressured formations inferred from seismic, thus can provide hints to possible hydrocarbon expulsion phase(s).

Secondary Migration

Secondary migration is the process of hydrocarbon moving to reach reservoirs. This requires ways of connecting the source with reservoir, known as pathways, a prime requisite in accumulation process. After expulsion from source, hydrocarbon continues to move up-dip, from high to low pressure areas through porous and permeable rocks until it reaches the structurally highest part of a reservoir, where it gets trapped and accumulated. This is also referred as charging of a trap which is a reservoir surrounded by nonpermeable rocks to prevent escape of hydrocarbon. Hydrocarbon migration continues to move and charge other traps present in the area depending on the connectivity by suitable pathways and the amount of hydrocarbon expelled. Expulsions from the source, however, may be upward or downward into the underlying formations depending on migration pathways and their closeness to traps located in the corridor (England and Fleet 1991). Downward migrations are usually short distance migrations and where the source overlies directly on the reservoir.

Migration Pathways

Migration pathways besides being the permeable formations, can also be facilitated by unconformities and faults which are considered critical elements in process of migration. Comprehending migration pathways for accumulation in traps requires knowledge of carrier (permeable) bed geometries with vertical and lateral permeability properties and the paleodips at the time of hydrocarbon expulsion. Migration paths are not simple two dimensional paths as is often assumed by interpreters, mostly perceived through faults. Migration is a more intricate three dimensional process which is challenging to predict precisely though can be analysed better by '3D basin modelling' (discussed later in the chapter). Contingent to proximity of source and reservoir, migrations pathways can be of long or short distance. Nonetheless, the pathways can be predicted broadly from seismic sequence and seismo-tectonic studies and by paleostructural analysis to indicate dip gradients, unconformities and faults. In simple systems, such as in a delta sequence where the porous delta-front reservoir sands are in contact with the contiguous pro-delta or marine source rocks, migration and charging processes are rather simple, short and straight forward as hydrocarbon expulsion charges the immediately juxtaposed traps. Vertically stacked delta sequences and growth faults with associated roll-over structures, are considered highly potential exploration plays that can be recognized and evaluated from seismic stratigraphy

and seismo-tectonics studies. Growth fault planes often considered as good conduits and the juxtaposed associated roll-over structural traps are ideal for hydrocarbon accumulation.

Migration Timing

Timing of secondary migration with respect to the presence of a trap is a key factor in the hydrocarbon accumulation process. Timing of migration and trap formation are critical as traps formed later than migration would miss hydrocarbon charging. It is thus essential the traps exist earlier to time of migration. Consequently, a valid structural trap may not have accumulation to show if the presently trap morphology, as mapped, happens to be altered by tectonics after hydrocarbon migration. While timing of expulsion and migration are determined precisely from thermal history by geochemical analysis, seismo-tectonic studies can broadly indicate the timing (age) of the structure (trap) formation. The analysis can ascertain the trap was existing prior to migration and that no tectono-morphological change occurred based on paleostructural and tectonic analysis. Since hydrocarbon moves up dip, it is also important to analyse the paleodips at the time of migration, together with presence of traps, to evaluate migration pathways and accumulation. Combined study of seismo-tectonic and seismic stratigraphy reveal chronological history and make it convenient to link the 'timing' factor for evaluation.

Entrapment and Preservation

Entrapment

Hydrocarbon entrapment occurs in a variety of traps—structural, stratigraphic and strat-structural (combination). Traps are reservoirs with lateral and top seals with nonpermeable rocks such as shales, tight limestones and evaporites (salt). Certain type of faults under specific conditions also act as seals. The trapping mechanism, however, varies and needs assessment of quality of seal and entrapment conditions for their integrity to act as effective traps to hold hydrocarbon. While structural closures are generally considered as safe traps, stratigraphic and combination traps can be of several types meriting careful analysis of their entrapment processes. Stratigraphic traps with updip lateral seal and fault-linked structural closures which are more common can be relatively more risky and may warrant special attention. This is because faults play significant role in migration, sealing, accumulation, distribution and preservation of oil and gas, discussed later.

Preservation

Preservation of accumulated hydrocarbon is an important component of discovering oil and gas and includes issues such as biological degradations, remigration and escape from traps. Poor preservation may be caused due to physical and chemical transformations through tectonic deformations, basinal tilts, gravity, erosions and diagenesis and also because of leaky seals. Remigration is mostly movement of hydrocarbon to structurally updip shallower reservoirs caused by tectonic events, particularly by reactivated and young faults post-accumulation that can play major role in redistribution of hydrocarbon. Detailed analysis of faults, their genesis and evolution history studied from seismo-tectonics studies can help evaluate preservation of accumulated hydrocarbon. However, occurrence of oil, gas and water in different layers in a thick reservoir formation with no discernible faults, obscures understanding the mode of such preferential accumulation of different phases of hydrocarbon with varying fluid contacts.

Hydrocarbon escaping to younger and shallower reservoirs from the main accumulation can sometimes be misconstrued as an element of new stratigraphic play for pursuing future exploration. Uplifts and erosions can biodegrade the entrapped oil by bacterial reaction and allow gas to escape resulting in heavy oil which becomes difficult and expensive to produce. Seismic sequence and seismo-tectonic studies can provide useful information on unconformities, faults and timing of traps with age to assist proper evaluation of trap integrity for entrapment and efficacy of preservation.

Basin and Petroleum System Modelling (BPSM)

Basin evaluation is usually carried out in an early stage of exploration to understand the hydrocarbon habitats in a basin and broadly estimate its resource potential. Resources and reserves are defined in various terms and types in the industry. Most simply stated, resources are assessment of amounts of hydrocarbon believed to be in the basin based primarily on amount and quality of the source rock and is different from reserves whish are proven quantities found in a reservoir. With the advent of sophisticated softwares and fast computers, synthesis of seismic, petrophysical and geochemical data to build unified geological model of the basin and its petroleum systems, is a common practice. The comprehensive modelling on a basin scale is called basin and petroleum system modelling (BPSM).

The basin and petroleum system modelling, essentially, aims to reconstruct the dynamic process of hydrocarbon sequence chronologically, from source and generation to accumulation and preservation. Each of the elements that is, source, temperature and pressure history, generation, evolution of trap, migration timing, accumulation, and retention are considered in the process to recreate the hydrocarbon sequence back in geologic time. The integrated basin and petroleum system modelling (BSPM) makes use of geological, geophysical, petrophysical and geochemical data. The geologic parameters mainly constitute the key tectono-stratigraphic frame work interpreted from seismic and well data, on which the predictive reliability of the modelling depends. The petroleum system modelling for hydrocarbon generation is essentially based on source amount and quality inferred from seismic and the thermal maturity history from geochemical inputs based on laboratory analysis of well samples along with other petrophysical and petrographical data. The next phase of migration, entrapment and preservation primarily dwells on the tectono-stratigraphic framework provided by geologic and seismic data.The modelling can be prepared in 2D or 3D mode depending on exploration requirements. However, the migration process being a complex three dimensional problem, is best realized by 3D basin modelling for applications in matured basins.

Petroleum System Modelling (PSM)

A petroleum system comprises a pod of an active source rock in a basin, genetically linked, to derived hydrocarbons, established by geochemical correlation (Al-Hajeri et. al. 2009).More than one source rock of similar nature but of different age, forms two or multiple petroleum systems. Petroleum systems are essentially controlled by the tectono-sedimentary 'kernel' of a basin and may have distinctive source and migration process characteristic of the basin genesis. For instance, petroleum systems in rift basins are likely to favor good source generation, reservoir facies and vertical migration paths through faults and fractures for hydrocarbon accumulation. Petroleum systems on inland platforms, on the other hand, may be relatively less favored due to abundance of reservoirs, but less of source rocks and with long, lateral migration paths (Perrodon 1992).However, appropriate timing of development and spatial disposition of the traps with respect to source kitchen are most critical for a petroleum system to happen.

Geochemical Evaluation of Source and Generation Potential

Elements responsible for hydrocarbon generation and accumulation have been discussed earlier in basin evaluation. However, the type of TOC in source and its quality and intensity of thermal maturity in a basin are the prime factors that decide the type and amount of hydrocarbon generated and need more deliberation on geochemical evaluation.

TOC and kerogen
Source rocks contain organic matters which under sufficient heat and pressure generate

hydrocarbon. The quality of source depends on the amount of organic content, total carbon content (TOC) and its type (Kerogen). Shale is the most common source rock and shows organic content varying usually from 1 to 10%, depending on their depositional environment. Nonmarine and lacustrine shales with higher organic content are considered rich sources, which constitutes at least 50% of organic content as total organic carbon (TOC, 0.5–5%).

Kerogen is an organic compound present in TOC which breaks down to hydrocarbon under temperature and pressure. Kerogen type varies in source rocks depending on their depositional environment and determines the amount and type of hydrocarbon produced. Kerogens are mainly classified under four types. Kerogen of type I are found in lacustrine sources and yields oil in plenty. Type II kerogen in deltaic to marine source rocks produce moderate amount of both oil and gas while type III belonging to terrestrial source is known to produce small amount of gas. While type I and II kerogens are formed from algal and marine bacterial remains under anoxic conditions, the type III kerogen is formed from decomposition of plant materials by bacteria under oxic conditions. The kerogen type IV contains inert organic matter and has little generation potential. Combination of kerogen types also occur in many situations with mixed sources as influx from different environment. The transformation of kerogen to hydrocarbon under increasing heat during burial starts with oil and then progresses to wet gas and finally to dry gas due to higher temperatures at deeper depths. Generation potential of kerogen types in terms of relative amount and type of hydrocarbon, oil/gas generated by pyrolysis (described later) is shown in Fig. 1.

Thermal Maturity
Thermal maturity is the key factor responsible in breaking the kerogen to generate hydrocarbon under heat and pressure and is an ongoing process as the organic matter undergoes burial under sedimentation. The degree of thermal maturity depends primarily on the paleotemperature and its gradient in the basin which increases with

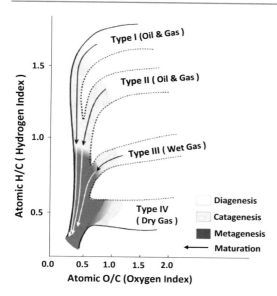

Fig. 1 Graphic showing type of hydrocarbon generated by the kerogen types under varying thermal maturity. H/C and O/C represent amount of hydrogen and oxygen released per carbon content in sample under heating (pyrolysis). Arrows indicate increasing maturity leading to different maturity stages. Note that type IV produces more oxygen with little hydrogen resulting in poor hydrocarbon generation (modified after Walters 2007)

burial depth. Thermal maturation is divided into three stages known as diagenesis, catagenesis and metagenesis based on degree of maturity which also controls the type of hydrocarbon product. The degree of thermal maturity is usually indicated by vitrinite reflectance R_o. Vitrinite reflectance and pyrolysis, also used as hydrocarbon generation indicator, are described later.

Diagenesis (% R_o <0.5) is a process of compaction starting under shallow burial when temperature and pressure conditions are mild. However, near surface chemical and microbial activities enrich the organic matters with more carbon content which on further burial breaks up to produce gas. This happens in a depth interval of order of a few hundred meters with temperature below 60^0 C. The gas is produced is small quantity and is mostly dry gas, methane, also known as biogenic gas.

Catagenesis (% R_o 0.5–2.0) ensues with deeper burial depth resulting in increase in temperature and pressure sufficient to crack kerogen into oil and then to gas. The process is called thermogenic and happens in the depth ranges of 1–3.0 km which is known as the 'oil window'. In the later part of catagenesis with further increase in burial depth, around 3.0–3.5 km, the generated oil is converted to gas known as 'wet gases'.

Metagenesis (% R_o 2.0–4.0) stage starts beyond depth of around 3.5 km and with increased depth of burial and attendant temperature and pressure, the kerogen produces what is known as thermogenic 'dry gases'.

Pyrolysis Estimate of quantity, quality and thermal maturity of organic matter in source rock is the most critical aspect of source potential evaluation, as it defines the different stages of type of hydrocarbon generation and their amount. This is typically determined in laboratory through a process called pyrolysis. Pyrolysis indicates thermal decomposition of organic matter when heated in absence of oxygen which produces hydrogen, oxygen and water from the sample. Measuring the quantity of hydrogen (H/C) and oxygen (O/C), also known as hydrogen index (HI) and oxygen index (OI), proportioned to carbon in the sample, signifies the quality of organic carbon and temperature at which thermal maturity is attained (refer Fig. 1). The Rock–Eval pyrolysis is the most common technique and the hydrogen index (HI) and the temperature index (T_{max}), the temperature at which maximum hydrocarbon is generated, are widely used as indices for thermal maturation of organic matter.

Vitrinite reflectance is another indicator of thermal maturity, usually expressed in units of percentage V_R or R_o. The R_o index measures percentage of light reflected from macerals which are intrinsic components of organic matter in the sample.

A composite graphic of thermal maturity intensity, vitrinite reflectance *(Ro)*, the pyrolysis temperature index (T_{max}), burial depths and the oil and gas window is shown in Fig. 2. Vitrinite reflectance (R_o) and pyrolysis methods *(T_{max})* and *(HI)* are used as indicators of hydrocarbon generation and characterize the kerogen types. However, ambient thermal gradients and pressure as well as source characteristics vary from basin to basin and also with geologic age and the Fig. 2 may be considered as typical representative indices. Based on thermal maturity values, source rocks are categorized as thermally immature, mature or post-mature signifying source capability and efficiency for hydrocarbon productivity.

Petroleum system modelling enables to assess the type of hydrocarbon and their relative abundance which makes its application gainful even in matured basins to find more hydrocarbons with lesser risk. Often drilling on a valid structural trap, contiguous to a producing oil or gas field, does not find success. Modelling can help find the reasons and guide to appropriate locales, predicted with more likelihood of success.

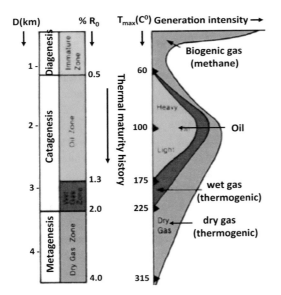

Fig. 2 Graphic of a composite display illustrating the stages of thermal maturity and type of hydrocarbon generated. The major elements that control the process, the burial depth, temperature, maturity indices of vitrinite reflectance (R_0) and T_{max} values are also indicated. (modified after Crain; depth and Ro added by author)

Limitations of Petroleum System Modelling

Highly sophisticated 3D basin modelling softwares are capable of addressing the complicacies of modelling petroleum systems including simulation of hydrocarbon expulsion and accumulation. Nevertheless, like in all techniques, the modelling can sometimes be flawed depending on the exactness of data input. The tectono-stratigraphic framework input, the main stay of petroleum system modelling is a subjective interpretation of data which to a large extent depends on the perception of the geoscientists. The geochemical inputs, measured in laboratories, could sometimes be misleading due to quality of samples used. Core samples from the well may not be retrieved fully or improperly processed and handled in the laboratory for measurement of geochemical parameters to provide dependable results. For instance, 3D modelling of an area with abundance of reservoir rocks and regional seals may predict a significant quantity of hydrocarbon generation and expulsion, yet subsequent drilling of several wells may establish no sizable accumulation. The drilling results may prove the elements of the petroleum system that went into modelling as legitimate except perhaps for the key element, the critical timing which is the lack of synchronization between timing of the peak expulsion and that of the deposition of regional cap rock to make the trap ready for accumulation. By far the most challenging part in the petroleum system modelling is about knowing the migration paths which can adequately explain the preferential accumulation of hydrocarbon in selected reservoirs in a corridor with others remaining uncharged.

Fault Attributes Analysis and Fault Seal Integrity

Faults in petroleum system play important roles as conduits, seals, and leaks in the process of migration, accumulation, and preservation and (re) distribution of hydrocarbons and form an integral part of Basin petroleum system modelling work flow. Fault identification and mapping is not an end to interpretation and herein comes the need to evaluate their bearings on petroleum system modelling and evaluation of a prospect before drilling. Fault, *per-se*, is usually neither a conduit nor barrier for migration and a seal for entrapment as customarily inferred from seismic sections and maps. Faults displayed in sections and maps are merely lines of discontinuity which indicate change in lithology, depth and dips of strata across it (Downey 1990).Critical analysis of fault attributes like type, genesis, displacement, age and history from seismotectonic studies, not only allow proper delineation and mapping but scope to analyse their impact on exploration ventures. Study of role(s) played by faults also can have significant effect on production and overall development of oil and gas fields.

Faults as Conduits

Extensional faults are in general considered more likely to enable hydrocarbon migration than compressional reverse faults. Growth fault planes are more likely to be conduits, particularly at shallow depths where they may behave as tensile and open fractures. At greater depths, where the growth faults are in overpressured zones, fluids driven by buoyancy moves up-dip along the fault plane resulting in short distance migration. The co-joined permeable zones, present along the fault planes, facilitate migration from source to reservoirs (Downey 1990) and is shown schematically in Fig. 3. Reverse faults caused due to shortening of strata in compressional regime have fault planes are less likely to be open to facilitate migration.

Faults as Seals

Hydrocarbon entrapment in structural traps formed against faults, known as fault –closure structures are typically envisaged due to the fault plane behaving as a barrier. The entrapment mechanism, however, depends on the type of permeable/nonpermeable strata juxtaposed across the fault, controlled by its amount of displacement (throw) which provides the lateral seal. This strictly needs mapping of the fault geometry

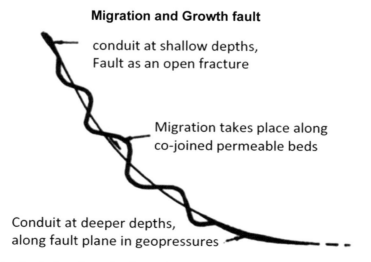

Migration and Growth fault

conduit at shallow depths,
Fault as an open fracture

Migration takes place along
co-joined permeable beds

Conduit at deeper depths,
along fault plane in geopressures

Fig. 3 A sketch showing hydrocarbon migration pathway along growth- fault. At shallow depths the fault behaves as open fractures and acts as conduit. At greater depths, growth faults, usually in overpressured regimes, promote upward movement of hydrocarbon driven by buoyancy of fluid. The fault plane co-joined with permeable zones present across the fault facilitates migration. (After Downey 1990)

in three dimensions for proper analysis. However, for easy illustration a 2D profile view of fault may be considered (Fig. 4). It shows that a lateral seal is formed by the fault when the throw juxtaposes impermeable beds against permeable (reservoir) beds (Fig. 4a), whereas, disposition of part of permeable beds across the fault leads to escape of some hydrocarbon (Fig. 4b). The disposition of permeable and non-permeable rocks across the fault is determined by the magnitude of fault throw vis-a vis the reservoir thickness.It is as an effective up-dip seal to form a trap for the whole reservoir only where the fault throw is more than the reservoir thickness (Fig. 4a) and a partial trap of limited thickness when throw is less than the reservoir thickness (Fig. 4b). However, the faults, under certain geologic situations may also behave as seals due to fault plane gouges (describer later) resulting in separate hydrocarbon contacts in the upper reservoir on either side of the fault caused by post-migration reactivated fault (Fig. 4c).

However, faults commonly change throws along their strike which can complicate the effective sealing process, especially in thin and multi-pay reservoirs. It is for this reason that the throws across faults need to be correctly estimated and represented by contours in the structure map (see Chapter "Seismic Interpretation Methods"). An elaborate approach to understand hydrocarbon migration and accumulation in fault-associated traps is by using an analytical technique known as 'fault-plane mapping' (Allan 1989).The analysis consists of mapping of strata juxtaposed across the faults three-dimensionally with appropriate structural dips to bring out the contact zones of permeable and nonpermeable rocks to judge the role of faults.

Fault Seal Integrity

Faults, in some instances can be considered seals where the fault plane behaves as a barrier. This can happen when impermeable rocks such as clays or cementing materials fill the fault plane zone as would be more likely in clay prone areas and old faults. Fault zones, especially of growth faults, developed in clastic sequences of predominantly shales can cause smearing of the fault walls with clay that separates the permeable strata on either side, a process termed "clay smear potential "(CSP) of fault (Doughty 2003). The behaviour of clay smears during the growth of a fault has important implications for fault seal analysis as the breaks in continuity can cause leakage of hydrocarbon. This, however, is hard to comprehend from seismic data and quantitative fault seal prediction in the subsurface becomes a challenging task (Doughty 2003).

Fault seal properties

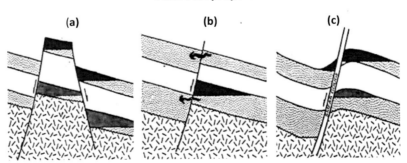

Fig. 4 Sketches illustrating fault properties. (**a**) Faults act as seal when fault throws cause impermeable rock juxtapose the permeable reservoir (most common), (**b**) faults behave as partial/non seals causing hydrocarbons to migrate partially or completely depending on throw that brings part or full disposition of permeable rocks against the reservoir thickness. (**c**) fault plane act as seals in the lower reservoir due to fault plane gouges. Interesting to note the separate fluid contacts in the upper reservoir in the blocks across the fault. This is because of the post-entrapment reactivation of the fault (Modified after Harding and Tuminas 1989)

Similar fault plane seals can be caused by granulations in the fault surface due to intense tectonic stress and cemented by diagenesis over a period of time. Big granules along a fault plane are known as 'fault breccia's' whereas, the smaller ones are known as gouges. Presence of shale in the fault plane binds the gouges to behave as a seal and quantitative analysis to estimate net shale content in a fault plane is useful in determining the efficacy of fault seal and is termed "Shale Gouge Ratio" (SGR). Though softwares are available to analyse sealing efficacy of clay smear potential (CSP) as well as shale gouge ratio (SGR) of faults, it is usually subjective and need calibration with subsurface data such as formation pressures and fluid contacts for successful predictions (Doughty 2003).

Occurrence of young or reactivation of old faults post oil and gas accumulations, may cause redistribution of trapped hydrocarbons in the reservoir, and in worst case scenarios, cause leakage of the entire volume of hydrocarbon. Structures with young faults, suspected for probable hydrocarbon leakage are called 'breached structures' and need caution for exploring. A schematic diagram illustrating the fault role (Fig. 5) and a seismic image (Fig. 6) of a 'breached structure' are shown. Fault attributes interpreted from seismo-tectonic studies assist in proper reconstruction of episodes of stress and deformation history to assess probable timing of hydrocarbon charge and accumulation. A cursory interpretation of a growth fault as a seal affecting the entire sequence, on close examination in seismic may reveal more information, the details of episodic history of a cyclic growth punctuated by intermittent reactivations. This would signify the discontinuities in the growth which, as mentioned earlier, can lead to leakage. It is thus vital to understand the genesis and age of faults affecting a structure, with respect to the timings of migration, as well as its status in relation to the structure as pre-existing, concurrent or post-charge for considering accumulation and preservation of hydrocarbon. This is particularly important if there are two or more sets of fault occurring at different times and without proper fault attribute analysis one may be led to flawed prospect evaluation resulting in exploration setback.

Fault analysis is also important during development and production phases as faults may impede flow continuity within the reservoir, leading to compartmentalisation of the field. Separate fault compartments may have separate fluid contacts which would require more wells to be drilled on the prospect, increasing the operational cost. Faults may also cause associated fractures, which introduce anisotropy (HTI) in the reservoir impacting production and water injection strategies.

Leaky Fault seals

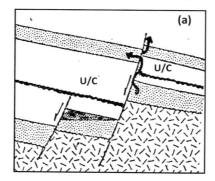 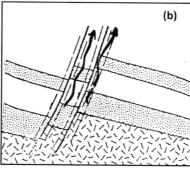

Fig. 5 Sketches illustrating role of faults occurring after hydrocarbon accumulation. Reactivation or generation of new faults (**a**) redistribute the trapped oil/gas in shallower reservoirs or (**b**) facilitate complete escape of accumulated hydrocarbon, leaking through fault planes. (Modified after Harding and Tuminas 1989)

Fig. 6 Seismic example of a 'breached' structure. Faults occurring after accumulation may facilitate escape of trapped hydrocarbon. *(Image: courtesy ONGC, India)*

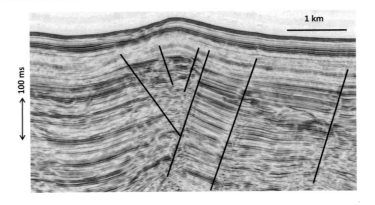

Prospect Generation and Evaluation

The aim of seismic stratigraphy and seismo-tectonics studies for basin evaluation is to identify geologic plays, leads and generate prospects for exploration. As stated earlier, prospects are those which are properly mapped and meeting all the elements required for hydrocarbon accumulation and await drilling. However, the prospects need technical evaluation before taking drilling decision as drilling involves hiring rigs, land acquisition, civil work and other infrastructures which require systematic planning and attendant expenditure. Prospect evaluation is often done by a simple and quick method known as probability of success. The three major elements of hydrocarbon accumulation having risk factors are the probabilities of (i) source and charging, (ii) good reservoir and (iii) trap mechanism which are considered for probability assessment for success. Risk is interestingly defined as quantifying the uncertainty and each of the aforesaid factors is assigned a risk ranging from zero to one depending on the analyst's judgement based on seismo-geologic studies. The risk-factor figures for potential drillable prospects are commonly around 0.7–0.8 for each and are multiplied to indicate the probability success of the prospect. Usually, a product figure around 0.4 and above is considered worth taking a chance to drill the prospect.

Techno- Economic Analysis

Techno-economics involves a more comprehensive risk analysis method for assessing their probable commercial values before making major drilling decisions. Evaluation deals with technical (geological) and financial risk analysis, an exercise which proffers an estimate of profitability of the exploration venture in case of hydrocarbon discovery. Evidently, the output of the exercise depends to a large extent on reliability and probability of the geological parameters which are input from seismic analyses, as well as on other important external factors such as commercial, political, logistical and environmental. We, however, restrict discussions to technical (geological) risk assessment.

Technical evaluation of a prospect is essentially based on the estimation of hydrocarbon volume in place and the quantity that can be produced from it. The parameters for volumetric estimate include the aerial extent and structural amplitude (thickness) of the prospect, the quality of source and reservoir rocks, and the type of traps with entrapment mechanism for an effective seal. However, a more significant point in the exercise, as discussed earlier, may be the assessment of the moment of trap formation with respect to the timing of hydrocarbon charge for accumulation and the post migration tectonics impacting preservation. The types of traps as stated earlier can be structural, stratigraphic and strati-structural—a combination type each with typical trapping mechanisms. The prospective traps commonly include four-way and fault closures, reservoirs with updip lateral seals, thin sands encased within shales and porosity pods within limestone and unconformity related traps. Each variety of trap warrants evaluation of its kind in terms of entrapment mechanism, quality of reservoir, and its extent and hydrocarbon column related to estimation of volumetric reserves.

Some of the critical geological parameters which play major part in risk evaluation linked intimately to financial consequences and drilling decisions are outlined. A few of these may be pertinent to later phase of development and production and may not be of immediate concern at early or intermediate stages of exploration. Yet the evaluation of these factors to assess attendant risks may be made as it helps the management to be aware of likely imminent exploratory inputs and related financial implications involved in near future. These risks are touched pointwise again to underscore the need of technical risk evaluation in a wider perspective for deciding future commercial and strategic plans in case a discovery is made. This is particularly applicable in situations where acreages and assets, offered on contract for exploration and development license, are to be geologically evaluated for risks before taking important decisions to bid for entering into exploration agreements.

Technical (Geologic) Risk Assessment

Type of Source

Forecasting the type of hydrocarbon find in the prospect, depends largely on the source type and is important because the engineering processes, infrastructures and economics to produce commercially gas and oil differ greatly depending on market, price and mode of transport etc.,. For a given prospect size, generally oil fields may need drilling of more production wells compared to that needed for a gas field. The capital and operational costs on engineering and infrastructures for developing a field, varies greatly with type of hydrocarbon and the field location, onland or offshore and the exploration strategy driven by resources of a company.

Migration Timing and Pathways

Source-reservoir connectivity and existence of trap at the time of migration is a prerequisite for accumulation which needs careful evaluation. In this context, presence of carrier beds, unconformity surfaces and faults acting as migration pathways, need to be carefully assessed more for their uncertainty in risk analysis.

Type of Reservoir (Clastic/carbonate)

Clastic and carbonate reservoirs often vary with different development and production plans. The productivity and recovery varies with the type of reservoir, e.g. sand reservoirs generally have higher productivity and higher primary recovery of in-place reserves compared to carbonates. Carbonate reservoirs, are more heterogeneous and complicated in nature, often fractured and offer relatively lower primary recovery. Secondary recovery processes for augmenting production can be different and vary in efficiencies for the two types of reservoirs. For instance, oil saturated sand reservoirs are generally more amenable to enhanced recovery through water injections whereas it may add problems in limestone reservoirs. Water flooding carbonate reservoirs can react with rocks and produce hydrogen sulphide gas (H_2S) along with hydrocarbon. The change of sweet oil (without H_2S) to sour oil (with H_2S) may require change in production engineering infrastructure of unforeseen high cost as H2S corrodes pipe lines during transportation and other attendant problems at the refineries.

Type of Trap

Traps, structural, stratigraphic or combinations may implicitly be risked accordingly. For instance, in a typical reservoir of structural pro-spect, the hydrocarbon rock volume estimate may be straightforward by multiplying structural closure (area) with its amplitude(reservoir thickness) but may not be so in a multi-pay stratigraphic thin reservoirs within the structural high. This is because of the uncertainty in assessing the number of the multiple thin reser-voirs, their varying thickness and lateral extent and more critically their distribution within the prospect. Purely stratigraphic traps may also require stricter assessment of trapping mecha-nism of lateral seals and may be considered more risk prone and accordingly factored in evalua-tion. Structural prospect with several criss-crossing faults may also need to be appropri-ately risked because of possibility of creating separate compartments. Each of the compartment may warrant drilling thereby increasing number of production wells and consequently the overall project cost. Faults may also be responsible for anomalous fluid flow patterns during production (Chapter Evaluation of High-Resolution 3D and 4D Seismic Data"), and prospects affected with a slew of faults may increase the prospective risk.

Estimate of Hydrocarbon In-Place (Volumetric)

Estimate of in-place hydrocarbon volume, also known as *volumetric* is the most important output of geoscientific evaluation and eagerly awaited by the management to work out financial calcu-lations for assessing cash flow. Hydrocarbon pore volume is estimated by multiplying the likely hydrocarbon bearing area of the prospect with thickness, porosity, and hydrocarbon satu-ration expressed by

$$V = A.h.\emptyset.S_h,$$

where
'V' is the hydrocarbon bearing rock volume, 'A' the prospect area, 'h' the hydrocarbon col-umn, 'ø' the reservoir porosity in percentage and 'S$_h$' the hydrocarbon saturation in percentage. In structural traps, where the hydrocarbon accrual

can be maximum up to the spill point of the structural closure, the volume parameters are usually computed optimistically by the contour-closed area (A) and amplitude of the structural closure (h). The reservoir porosity and hydro-carbon saturation are presumed geologically based on the type of likely reservoir, the porosity values presumed around 15- 20% and hydrocar-bon saturation of $\sim$ 75–80%. The volume of hydrocarbon available for recovery at the sur-face, however, is computed by multiplying the in-place reserve by a conversion factor. This is because oil and gas when brought to surface, they change volumes. It is known as the *shrinkage* factor for oil as the volume at surface is reduced due to escape of dissolved gas from the oil produced at surface. Though the factor depends on the oil gravity, amount of gas dis-solved in oil, temperature and pressure, stated empirically, lighter the oil more is the shrinkage factor. For gas, however, it is the reverse way, gas expands in volume at surface and the factor is known as gas formation volume factor (FVF). Since the estimates are based on reservoir parameters assessed from interpreted seismic maps and geologic prediction of petrophysical parameters all these parameters have inherent uncertainties and are likely to be highly influ-enced by subjectivity laced with analyst's indi-vidual bias. Skill and experience of an interpreter can help come out with a judicious realistic estimate of hydrocarbon volume in place.

Based on volumetric reserves, recoverable reserves and production profiles are calculated and economic analysis is carried out.The techno economics are considered feasible depending on reasonable rate of return on investment (ROI) and decide future exploratory inputs including number of wells to be drilled to develop the prospect. In case of a discovery, exploration enters into the appraisal phase, in which additional exploratory inputs are acquired that include detailed seismic and drilling of more wells generally known as delineation and appraisal wells to determine commerciality of production. Fresh analyses are mostly done in detail to reduce risks before committing large investments to develop the prospect as a field for production. Techno-economic risk analysis underscores the need for reliable seismic interpretation and the importance of seismic maps prepared on which reserve estimates are carried out and further exploration strategies are drawn. It may be also stressed that seismic stratigraphic and seismo-tectonics studies continue to provide sensible information not only at early phases but at all phases of exploration including production.

References

Al-Hajeri MM, Saeed M et al. (2009) Basin and petroleum system modelling. Oilfield Rev 21(2)

Allan SU (1989) Model for hydrocarbon migration and entrapment within faulted structures. AAPG Bull 73:803–811

Crain R Crain's petrophysical hand book

Downey MW (1990) Faulting and hydrocarbon entrapment. Lead Edge 9:20–22

Doughty PT (2003) Clay smear seals and fault sealing potential of an exhumed growth fault. Rio Grande Rift, New Mexico, AAPG Bull 87:427–444

England WA, Fleet AJ (1991) Petroleum migration. Introduction Geol Soc London Spec Publ 59:1–6

Harding TP, Tuminas AC (1989) Structural interpretation of hydrocarbon traps sealed by basement normal faults at stable flanks of fore deep basins. AAPG Bull 73:812–840

Perrodon A (1992) Petroleum systems: models and applications. J Pet Geol. https://doi.org/10.1111/j.1747-5457.1992.tb00875

Walters C (2007) Chapter—the origin of petroleum. Prac Adv Pet Process. https://doi.org/10.1007/978-0-387-25789-1_2

Direct Hydrocarbon Indicators (DHI)

Abstract

Presence of gas, particularly in young porous sands significantly lowers the bulk modulus and typically creates high amplitude seismic anomalies known as 'bright spots', usually considered direct hydrocarbon indicators (DHI). Though high amplitudes are more commonly the characteristic of gas, light and normal grade oil can also manifest high seismic amplitude responses. However, all bright anomalies are not caused due to hydrocarbons and therefore need proper validation before drilling. Other amplitude anomalies such as 'dim' and 'flat spots' also indicate oil/gas reservoirs and may be considered as DHI anomalies. The genesis of DHI anomalies is illustrated with graphics and seismic images. Evidence of interval velocity, reflection polarity, and phenomena such as reflection 'shadow zone' and time 'sag' below the, oil/gas reservoirs manifested in seismic corroborate hydrocarbon bearing DHI anomalies and are included in discussion.

Validation of DHI anomalies are commonly done through rigorous AVO analysis on seismic 3D data (Chapter "Shear Wave Seismic, AVO and Vp/Vs Analysis"). However, an alternate simple and straight forward way to validate DHI amplitude anomalies can be by analysing angle- stack amplitudes. Reflection amplitude depends on angle of incidence and the property of angle dependent P-reflectivity at near and far-angle is conveniently utilized to indicate reservoir type and fluid content. The near-angle reflectivity shows the type of reservoir matrix whereas, the far angle the fluid. The quick-look method is useful to delineate and characterize much faster and is expounded with help of case examples of offshore hydrocarbon sands. DHI anomaly related shortcomings are also outlined.

Seismic reflection amplitudes are produced as a result of impedance contrast of rocks at interfaces that the seismic waves encounter in the subsurface. Impedance of sedimentary rocks depend on their composition, i.e., the matrix, the density and the pore fluid, each of which influences elasticity (Chapter "Seismic Wave and Rock-Fluid Properties"), mainly the compressibility (bulk modulus). Gas is highly compressible and when present in sedimentary rock-pores, appreciably lowers the bulk modulus making it more compliant than rocks saturated with oil or water. This results in causing a high P-impedance contrast and producing strong seismic amplitudes. The strong amplitudes are used as good discriminator of hydrocarbon and often makes it convenient for detection of gas and are accordingly termed 'direct hydrocarbon indicators' (*DHI*). However, it may be stressed that all high amplitude anomalies may not be due to the presence of gas in the reservoir and terming high amplitude anomalies as DHI may be inappropriate. The hydrocarbon bearing high-amplitude anomalies (*'bright spots'*) have their characteristic attributes and contingent to specific conditions and geologic settings. There are also other seismic responses that indicate hydrocarbon saturated reservoirs such as *'dim spots'* and *'flat spots'* but do not show strong amplitudes. Interval velocity, reflection polarity and phenomena of reflection *'shadow zones'* and *'sags'* can also be important indicators for gas and may be considered when validating DHI amplitude anomalies for hydrocarbons. Historically, though DHI high amplitude anomalies are linked to gas, these can also be caused by oil saturated rocks. However, the high amplitude gas anomalies are more conspicuous, common and occur widely compared to oil anomalies.

High amplitude anomalies (DHI), customarily attributed to gas saturated sands are characterized by distinct seismic properties described later. Typically, the high amplitudes are due to decrease in impedance of the hydrocarbon saturated reservoir with respect to the overlying cap, caused by substantial lowering of bulk modulus of both matrix and the pore-fluid. Nonetheless, the presence of light oil in rock pores, are reported to have manifested high seismic amplitudes similar to that of gas (Whang and Lellis 1988; Clark 1992; Bulloch 1999). This is believed caused by lowering of bulk modulus due to the large quantity of gas dissolved in light oil. However, studies made by others indicate to the contrary that no such effect is exhibited by gas dissolved in water (Gregory 1977; Wang 2001). Osif (1988) reported little to no effect of gas in solution on the compressibility of water or brine. Experimental measurements by Liu (1998), in contrast, showed a reverse trend, a slight increase in acoustic velocity of dissolved gas in water. Interestingly, Anstey (1977) had stated categorically that only free gas in pores can cause the lowering of bulk modulus to result high amplitudes and not by solution or dissolved gas. However, while it continues to be a contentious issue, normal grade oil in rock pores can also cause high amplitude anomalies that correlate to the oil sands tested in the wells in east coast, offshore, India (Nanda and Wason 2013).

Obviously, the high amplitude response of the normal grade oil saturated sands is primarily due to compressibility of the soft reservoir matrix, aided little by the pore fluid.

Rock and fluid properties such as elasticity, density, texture, porosity and pore shape, fluid type and saturation, viscosity and other factors, the pressure and temperature, all affect the elasticity and density of a rock in one way or the other, which consequently influence the seismic response (Chapter "Seismic Wave and Rock-Fluid Properties"). However, the rock matrix, the porosity and the pore fluid mainly influence the DHI amplitude responses. Fully saturated oil sand may show noticeable difference in seismic response from fully saturated water sand but the difference with partially saturated oil sand may be marginal to be discernible in seismic. Elastic and other properties of individual rock can vary differently and their cumulative effect can affect the resultant seismic properties in myriad ways. For instance, while porosity in a rock decreases velocity and bulk density, fully water saturated rocks tend to increase the seismic properties because of increased incompressibility. Some of the effects may be individually too small to be perceived in seismic, but a few of them when added together can create a discernible change in seismic. Predictably, the seismic responses would thus vary widely from region to region depending on geology of the area which fundamentally controls the rock and fluid properties. Varied depositional scenarios thus can have dramatic effect on the compressional velocities of rocks and consequently DHI signatures in seismic can differ greatly. Such responses have been modelled by Wander et al. (2007).

An interesting phenomenon that merits mention is the reflection caused not by lithological contrast but by contrasts of fluids present in a reservoir such as the gas-water, gas-oil and oil-water contacts. Such seismic expressions can often offer clues as DHIs. The seismic reflection is typically a flat (horizontal) event, known as 'flat spot', which is primarily caused by the contrasts in fluid density aided by the contrast in bulk modulus of the fluids saturated matrix at the contacts within reservoir and is discussed in detail later. Expectedly, 'flat spot' corresponding to gas-water (brine) interfaces, having significant fluid density contrast, are likely to be noticed more prominently than the oil-water contacts and particularly in thick reservoirs, as demonstrated by Schroon and Schüttenhelm (2003). Oil and water saturated rocks exhibit near similar acoustic impedance, and though oil-water contacts may be imaged by seismic under favourable conditions, they are rather infrequent. For instance, light oils with large quantities of dissolved gas under pressure in insitu reservoir condition can appreciably reduce the bulk density to cause contrasts at oil-brine contacts for producing perceptible 'flat spot' seismic anomalies.

DHI Amplitude Anomalies

Though high amplitude 'Direct hydrocarbon indicator' (DHI) anomaly, as the terminology suggests, is indicator of hydrocarbon, it is more commonly considered gas because of its seismic response, the strikingly strong amplitude anomaly easily identifiable. Nevertheless, as stated earlier, oil in certain situations can also show similar amplitudes and is deliberated later in the chapter. There are three types of DHI amplitude anomalies associated with gas reservoir which are seen in normal stack seismic sections, called 'bright', 'dim' and the 'flat spot'.

'Bright Spot'

'Bright spot' anomalies are characterized by strong reflection amplitude with negative reflection coefficient ($-ve$ R_C), and are caused by hydrocarbon bearing sands capped by shale. A simple representative geologic model illustrates this (Fig. 1). Considering the model, the water-saturated sand exhibits velocities and densities, (e.g., 2300 m/s, 2.2 g/cm^3) close to that of overlying shale (e.g. 2100 m/s, 2.3 g/cm^3). The marginal positive impedance contrast ($+R_C$) at top of sand would generate

weak-amplitude reflection with positive polarity. However, when the water is replaced by gas, the velocity is considerably reduced along with slight density decrease (e.g., to 1600 m/s and density 2.1 g/cm^3) thus causing a significant decrease in impedance with negative contrast at the interface. This results in a gas-sand-top reflection showing bright amplitudes and negative polarity and is called the 'bright spot', typical of gas-sands. Following the model in the schematic diagram (Fig. 1), the gas water contact would be represented by a moderate flat reflection with positive polarity ($+R_C$) in contrast to 'bright spot' reflection polarity. Down-dip, below the fluid contact, the shale- water sand interface along the flanks would exhibit weak reflection amplitude with positive polarity ($+R_C$). Polarity is thus the most important criteria to define 'bright spot' and requires proper identification (see Chapter "Seismic Reflection Principles—Basics"). A seismic expression of a 'bright spot' for a gas sand is shown in Fig. 2.

'Dim Spot'

'Dim spot' is a weak—amplitude reflection with a positive polarity ($+R_C$) linked mostly to gas in limestone and old-age sandstone reservoirs. We would consider again a representative geologic model to explain the anomaly (Fig. 3). Water saturated limestone with velocity and density (~ 3400 m/s, 2.4 g/cm^3), much higher than the cap shale (~ 2600 m/s, 2.3 cm^3) would cause reflections with strong amplitudes and positive polarity ($+ve$ R_C). Impregnated with gas, the velocity of the limestone reservoir would be greatly reduced with slight lowering of density (e.g. 2900 m/s, 2.3 g/cm^3) and result in decrease in impedance and the contrast. However, it may be noted that the though the impedance contrast is lowered it is still positive which causes reflections at carbonate reservoir top with weak amplitudes and positive polarity at the top of the structure. This is called a 'Dim spot' DHI anomaly. The reflection coefficient, however, continues to remain positive and much higher

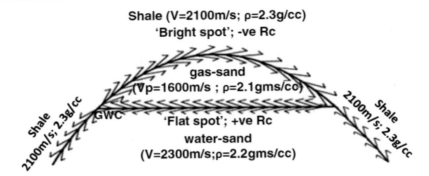

Shale (V=2100m/s; ρ=2.3g/cc)
'Bright spot'; -ve Rc

gas-sand
(Vp=1600m/s ; ρ=2.1gms/cc)

'Flat spot'; +ve Rc
water-sand
(V=2300m/s;ρ=2.2gms/cc)

Fig. 1 Representative geologic model of gas-sand Illustrating seismic amplitude anomaly 'Bright spot'. Gas sand top having large negative impedance contrast with overlying shale causes strong amplitudes with negative polarity at the crestal part. The reflection amplitude reverses to weak and positive polarity down-dip at flanks below gas-water contact where the contrast between shale and water sand is positive. Note the flat reflection shown with moderate amplitude and positive polarity due to contrasts in the fluids, the gas and water (flat spot). Arrows to left indicate trough (negative Rc) and to right the peak (positive Rc).

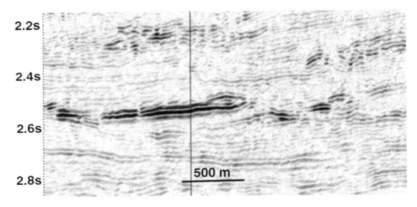

Fig. 2 A seismic segment showing an expression of DHI 'Bright Spot' amplitude anomaly. Strong amplitudes and negative polarity of the reflection from top of gas sand characterize Bright spots. Base of gas sand is imaged as high amplitude positive polarity due to contrast with water sand (Image: courtesy ONGC, India)

beyond the fluid contact and accordingly the flank reflections would show strong reflection amplitude contrasting the weak reflection at the top of gas limestone reservoir. A seismic 'Dim spot' is thus characterized by weak reflection amplitudes and positive polarity. Example of an offshore limestone gas reservoir is shown, where the lateral extent of gas reservoir can be mapped by the 'dim' crestal amplitude (Fig. 4). 'Dim spot' anomalies are also caused by old-age sandstone gas reservoirs having similar high-impedance as limestones reservoirs. The 'flat spot', similar to gas-water contact can also be seen in the model limestone reservoir (Fig. 3)

and can also hold good for sandstone gas/oil reservoirs. It may also be underscored that high amplitude anomalies at deeper depths are unlikely to be due to hydrocarbon bearing sandstones, which usually being of older age are more likely to exhibit 'dim spot' anomalies in seismic.

'Flat Spot'

'Flat spot' anomalies are characterized by moderate to high amplitude flat (horizontal) reflection events with positive polarity and are mostly associated with gas water contacts as mentioned

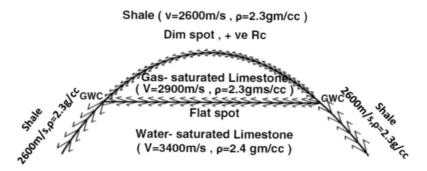

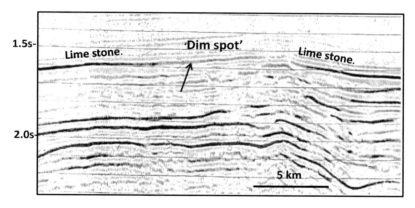

Fig. 3 Representative geologic model illustrating seismic 'Dim spot' anomaly in limestone. Gas in limestone lowers the positive impedance contrast with top shale and causes 'Dim spot', an amplitude anomaly characterized by weak and positive amplitude. Reflections from contact of shale and water saturated limestone, on flanks, down-dip to fluid contact, however, indicate positive polarity and much higher amplitudes compared to the crestal part. The 'Flat spot' anomaly is also shown at the fluid contact.

Fig. 4 Seismic segment showing 'Dim spot' anomaly. Gas in limestone lowers the positive impedance contrast with overlying shale and causes weaker amplitudes with positive polarity. Note the weakening of amplitude due to gas at the crestal part of the structure and the high amplitudes downdip on the flanks caused by shale and brine-saturated limestone contact. The 'dim spot' extent can be mapped to delineate the gas reservoir. Incidentally, no evidence of 'flat spot' indicating gas-water contact can be seen (Image: courtesy ONGC, India)

earlier (Figs. 1 and 3). The horizontality nature of the reflection signifies a fluid contact, gas water (GWC) or oil water (OWC) which are mostly horizontal though tilted oil-water contacts are also known to occur depending on hydro-dynamic conditions. 'Flat spot' reflection ($+ve$ Rc) polarity is characteristically opposite to that of 'bright spot' anomalies and seen together are considered more reliable indicator of hydrocarbon (Fig. 5). 'Flat spot' can occur in structural as well as stratigraphic traps and its disposition vis-à-vis the trap is important. For instance, an otherwise valid 'Flat spot' reflection event seen in a syncline or off structural highs cannot be a fluid contact but an artefact. Nonetheless, near horizontal event in a known gas prone area can sometimes be misinterpreted. One such flat reflection with strong amplitudes, inferred as a 'flat spot', on drilling turned out to be spurious, a water saturated sand (Fig. 6).

Amplitude anomalies are known to occur due to several other reasons than the rock and fluid properties such as lithology, reflector geometry, thin-bed tuning effects, propagation effects, interference of reflections, and noises. Consequently, these DHI anomalies may not always be

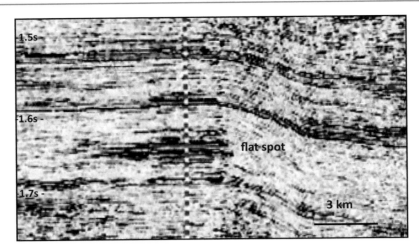

Fig. 5 Seismic segment showing 'Flat spot' anomaly . Flat spot is characterized by horizontal reflection event caused by fluid contacts due to density contrasts between oil/gas and water within a reservoir. Fluid contacts are imaged with positive polarity and amplitude depending on the degree of contrast in fluid densities. Note the disposition of the fluid contact, aptly located at the crest of the structure (Image: courtesy ONGC, India)

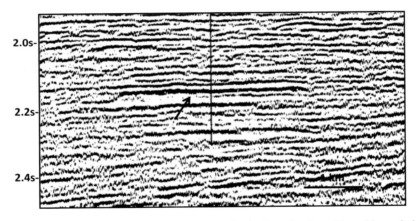

Fig. 6 Seismic example of a pseudo 'flat spot'. The high amplitude, flat reflection with positive polarity on drilling, turned out brine saturated Oligocene sand in offshore (Image: courtesy ONGC, India)

considered a decisive criterion to indicate hydrocarbon based only on amplitude. Other supportive criteria are therefore necessary to authenticate the DHI anomalies before drilling.

Supporting Evidence for DHI Anomalies

(i) *Polarity*

DHI anomalies, as stated earlier, are explicitly associated with distinct reflection coefficient criteria; 'bright spots' with negative polarity and 'dim' and 'flat spots' with positive polarity. The polarities thus are reliable indicators of reservoir lithology, fluid and contacts. A high amplitude reflection of negative polarity abruptly reversing to positive polarity can be a strong evidence of up-dip gas sand watering out down-dip (Fig. 7). Though, determining polarities of DHI amplitude anomalies is extremely crucial, in practice it at times may be difficult to precisely estimate the true polarity particularly for thin sands. Limited seismic band-width, interference of

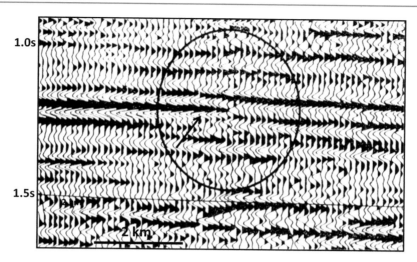

Fig. 7 Seismic segment showing example of polarity reversal associated with a DHI 'bright spot' anomaly due to change of fluid saturation downdip. Reflection from top of gas-sand with high amplitude and negative contrast (trough, white) reverses to positive polarity (peak) with amplitudes somewhat reduced with the gas-sand changing to water- sand having positive impedance contrast with shale (After Anstey 1977)

reflections, presence of noises and unknown polarity display convention affect estimation of true polarity.

(ii) *Velocity*

Significant lowering of interval velocity due to gas in a reservoir may be detected from seismic NMO (normal move out) velocities. Lowering of P-velocities combined with high amplitude of anomaly act as more reliable for detection of gas. However, conventional interval velocity computations require reflections from top and bottom of the reservoir. The top and bottom reflections in many instances may not be resolved if the reservoir is not thick enough or due to marginal impedance contrast at the reservoir top or bottom. For instance, for fining upward channel sands, the top contrast may be gradual and may not evince a reflection, even though the discrete bottom may show a high amplitude reflection. Similarly, for a bar sand, the top gas sand may be imaged well but not the bottom being a transitional reflector (refer Chapter "Seismic Reflection Principles—Basics") due to the gradual change in contrast. Furthermore, the computation of interval velocity from NMO is highly sensitive to interval thickness and can be often flawed, particularly in case of thin reservoirs.

(iii) *'Sag' Effect*

Velocity lowering due to gas in a reservoir can create artefacts, called as 'time anomalies' below the reservoir. In case of a thick gas reservoir, arrival times of reflectors below the reservoir will be time delayed due to lower velocity and indicate a local depression, vertically below the gas zone. This is called '*sag*' or '*pull-down*' effect and is exemplified by a geologic model, its computed response and actual seismic response (Fig. 8).

(iv) *Reflection Shadow Zones*

Low-frequency shadow zone observed below 'bright spots', are believed by many due to absorption of energy by the gas in the reservoir. The absorption phenomenon in gas has significant and wider geologic connotations in exploration applications. Technique based on energy absorption which is frequency selective is developed for direct detection of hydrocarbons by examining the amplitude spectra for decay of higher frequencies. This is known as 'sweetness' analysis, where absorption is mathematically

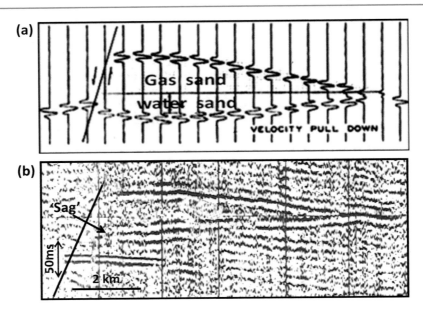

Fig. 8 Figure showing seismic modelling to validate a seismic bright amplitude DHI and the associated 'sag' feature seen in seismic. **a** computed seismic response of the gas-sand geologic model and **b** the field seismic. Note the 'pull-down' or 'sag' feature seen in synthetic (**a**) below the gas sand which matches with real seismic (**b**) The seismic interpretation of gas sand and related the 'sag' below the reservoir, the 'pull-down' artefact caused by low velocity due to gas, is validated by the modelling (After Anstey 1977)

determined as the amplitude divided by square of frequency. Another application of frequency absorption technique is for detecting faults as open or closed (Strecker et al. 2004). High energy (frequency) absorption by fault, studied from amplitude-frequency spectra in successive time windows, as a function of travel time, may signify open fault plane which help hydrocarbon migration and accumulation and increased permeability for production. Low energy attenuation, on the other hand, would imply cemented fault planes indicating sealing faults.

Absorption of energy in gas is a contentious issue, Absorption is frictional loss and gas being slushy, is unlikely to cause substantial friction for perceptible loss of energy (Gregory 1977; Anstey 1977). However, partially saturated hydrocarbon reservoirs are reported to show some degree of absorption loss because of relative motion of fluids against the pore walls within the rock but is doubtful if it is ample to be discernible in seismic. Without substantial absorption of energy in gas, the phenomenon of reflection shadow below gas reservoirs remains

debatable. Despite several explanations advocated, none seems good enough to explain reasonably the low-frequency shadow zones linked to gas reservoirs (Ebrom 2004). Extensive studies by Ebrom, nonetheless, have identified a few likely mechanisms for such phenomena based on numerical forward modelling. This includes study of type of reservoir rock matrix with high absorption factors and CDP stack-related problems during data processing that can lead to loss of high frequencies. Improper stacking of data with noise, poor deconvolution and static correction, inappropriately picked velocity and inefficient migration can also result in low frequency dominated reflections. Furthermore, shadows can also be caused due to transmission losses in a thick formation consisting of multiple thin gas reservoirs with alternating signage of impedance contrasts similar to absorption (see Chapter "Seismic Wave and Rock-Fluid Properties").

However, many thick gas reservoirs do not demonstrate low-frequency shadow phenomenon which may lead to surmise that absorption due to gas cannot be the sole reason for the phenomenon. Other factors such as the reservoir matrix, presence of multiple thin units, vertical heterogeneity in facies and interference of multiples from thin beds and poor data processing may account for the shadows.

Validation of DHI Anomalies

Angle Stack Amplitudes, Near- and Far- Stacks for Validation

Seismic high amplitude (DHI) anomalies picked on stack sections are normally authenticated by AVO and Poisson's ratio studies (Chapter "Shear Wave Seismic, AVO and Vp/Vs Analysis") before drilling for hydrocarbon. Such analysis needs quality 3D seismic data providing prestack NMO corrected gathers and shear velocity (*Vs*) data, which may not be always available. An alternate and simple way to validate DHI anomalies is the study of angle stack amplitude sections. The technique can be extremely effective not only in predicting hydrocarbon sands but in delineating and characterizing the hydrocarbon sands in the area.

Reflection amplitude depends on angle of incidence and the property of angle dependent P-reflectivity at near and far-angle is conveniently exploited to indicate reservoir type and fluid content. The near-angle reflectivity shows the reservoir matrix whereas, the far angle, the fluid. Seismic sections are usually stacks of all traces after transforming the offset distances into angles of incidence and exhibit average amplitudes. However, stacks for selected angles such as the near, median and far angles can be easily prepared and are known as 'angle stacks'. The near and far angle stacks are stacks of the near and far offset traces unlike the normal (full) angle stack section that includes all the traces. Evaluation of angle stack amplitudes is a simple and straightforward process involving visual comparison of amplitudes in the near and fare-stack sections but is highly convenient and effective as a tool for quick authentication of DHI anomalies. Reflected P-amplitudes are known to vary with angle of

incidence (offset) due to generation of mode converted shear wave. While generally, the reflection amplitudes decay with increasing offsets for nonreservoir and water saturated rocks, it shows the opposite effect for low-impedance hydrocarbon sands, that is, the negative amplitudes($-$ve R_c) may increase with increasing offset (Chapter "Shear Wave Seismic, AVO and Vp/Vs Analysis"). As a general rule of thumb, higher amplitudes seen in far- stack sections than in near-stacks are considered as indicators of hydrocarbon sands whereas strong amplitudes in near-stack but weakening in far- stacks, indicate otherwise, the nonreservoirs or water sands. The near, normal full and far stack sections, when compared, look remarkably different and makes it expedient to pick the strong amplitude anomalies seen in the far- stack sections only as potential hydrocarbon sands (Fig. 9). It may be noted that high amplitude anomalies seen in normal (full) stacks, are most likely to be non-hydrocarbon bearing if not seen in the far angle stack.

Delineation and Characterization of Hydrocarbon Sands

Essentially evaluation of angle-stack amplitudes is a comparison of near-normal (near angle) and far-angle P-impedances. While the amplitudes in near-angle stacks are influenced by normal impedance contrast, for a hydrocarbon bearing sand, the amplitude increase at far-angle stacks is influenced mainly by the lowering of Poisson's ratio (*Vp/Vs*) which involves shear wave velocity. This is expounded in detail along with mode converted P-wave to shear wave in Chapter "Shear Wave Seismic, AVO and Vp/Vs Analysis ". Thus while the near-angle stacks would denote the matrix (sand) the far-angle would indicate fluid content (hydrocarbon). The fact of amplitude variance with offset (angle stacks), from near to far stacks can be logically stretched to indicate not only the reservoir and pore fluid content but also their types, gas and oil. The quality of the reservoir and the type of hydrocarbon, gas or oil, can be predicted albeit

a) Near-stack   **b) Full-stack**   **c) Far-stack**

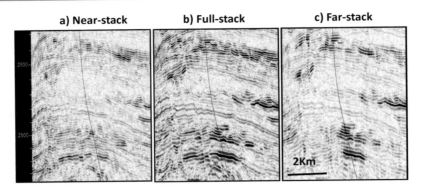

Fig. 9 Seismic segments showing comparative amplitude displays of **a)** near- , **b** full- and **c** far- angle (offsets) stack showing variations in amplitude anomalies. Bright amplitude in far-angle stack **c** indicates hydrocarbon whereas, high amplitude anomalies in full stack **b** but not seen in far-offsets are indicative of nonhydrocarbon bearing rocks. Bright amplitudes in far but not seen in near are potentially better hydrocarbon sands (refer next Fig. 10). Note the significant amplitude changes in far- stack **c** from normal full- stack **b** from which the DHI anomalies are initially picked as prospective anomalies (Image: courtesy, ONGC, India)

qualitatively depending on the geologic setup. Particularly, for offshore young Pliocene channel sands which are often the hydrocarbon habitats, it can be a simple and convenient way for preliminary delineation and characterization of hydrocarbon sands. The method leads to quick evaluation of in-place reserve and production potential to firm up the future exploration strategy. This is showcased below by a real case example dealing with oil and gas sands in the offshore area.

Case Example

The example is from offshore, east coast, India where Pliocene oil and gas sands have been discovered. Commonly, the offshore channel sands exhibit sand reservoirs of two qualities, the good clean channel sands and the inferior levees shaley-sands. Good reservoir sands usually have initial P-velocities slightly higher than the cap-shale, which after oil saturation would exhibit reduced velocity close to that of shale. Since the bulk density of the porous reservoir sand is usually lower than shale, the resultant impedance contrast tends to be small, showing no or poor amplitudes in near-angle stack (Fig. 10a). On the other hand, a shaley-sand reservoir with initial velocity close to that of shale, after oil-saturation would show considerable reduction in velocity and added with lowered bulk density would cause fair impedance contrast (negative) resulting in moderate amplitudes in near-stack (Fig. 11a). In both the cases the far-angle stack, though, would display strong amplitudes because of oil (Figs. 10b and 11b). It may be noted from these images that the amplitudes are more influenced by the type of sand matrix than by the fluid (oil) content in the near-angle stacks. However, the amplitude responses are different in case of for gas-sands. Free gas, being highly compressible, significantly lowers the sand velocity and the bulk density and causes high negative impedance contrast between cap shale and top of gas sand. This causes strong amplitudes in near-angle and stronger in far-angle stack as seen in Fig. 12a, b. Gas sands are thus likely to show high amplitudes ('bright spots') in both the angle stacks, the near- and far-angle, type of matrix (quality) playing a secondary role to pore fluid content (free gas). It can be surmised that oil sands though can produce high amplitude anomalies are unlikely to cause classical 'bright spots' which are, characteristic of gas sands (Nanda and Wason 2013). Evaluation of amplitude variances in near and far-angle stacks can thus help distinguish and delineate the oil and gas sands and also predict oil sand reservoir quality and hence its production

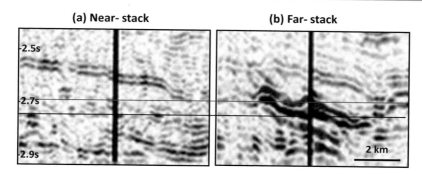

(a) Near- stack   **(b) Far- stack**

Fig. 10 The seismic example shows amplitude response of a good oils and reservoir in near- angle and far-angle stacks. Note the imperceptible amplitudes in near-angles stack that shows high amplitude in the far- angles stack signifying good quality oil sand. Poor amplitude in near- stack is due to small contrast in P-impedance as the initial higher (than shale) *Vp* of the good sand gets lowered due to oil saturation resulting in impedance similar to shale (After Nanda 2017)

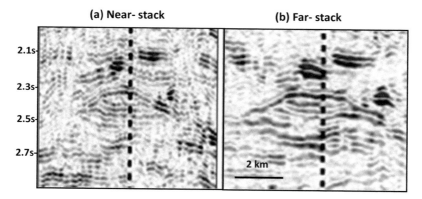

(a) Near- stack   **(b) Far- stack**

Fig. 11 Example showing seismic amplitude of a shaley oils and in **a** near- and **b** far-angles stacks. Note the moderate amplitude in near-stack which gets brighter in far-stack. The moderate amplitude is due to initial negative contrast in P-impedance ($\sim Vp$) of the shaleys and (compared to good sand) , getting considerably reduced after oil saturation and resulting in moderate contrast with shale (After Nanda 2017)

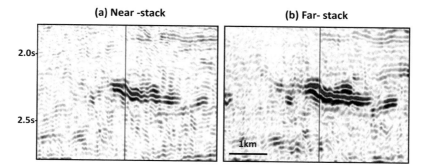

(a) Near -stack   **(b) Far- stack**

Fig. 12 Seismic example Illustrating amplitude response of a gas sand ('bright spot') showing strong amplitude in **a** near- and stronger in **b** in far-angles stack. The strong amplitude in near-angles stack is due to free gas in pores which significantly reduces the velocity and bulk density resulting in considerable negative impedance contrast to cause bright amplitude which becomes brighter in far-angles stack. Gas sands characteristically show strong amplitudes in both near and far angle stacks irrespective of type of sand matrix (After Nanda 2017)

potential (Nanda 2017). To summarize the empirical inferences -

- Feeble amplitudes in near-angle and high amplitudes in far-angle stack are indicative of good quality oil sand signifying better production potential.
- Moderate amplitude in near- and high amplitude in far-angle stack indicates shaley-oil sand and consequently of inferior production potential.
- Strong amplitudes at both near- and far- angle stacks ('bright spot') indicates gas irrespective of quality of the sand.

However, there are caveats to be aware of; seismic amplitudes are influenced by several other factors besides the reservoir matrix and fluid content and that the angle-stack amplitude studies described above are empirical, qualitative and geologic- area-specific. Depending on the variations in rock-fluid properties expected in diverse depositional settings, the amplitude patterns can vary in many ways, and add uncertainties to prediction and is exemplified by case studies in east coast, India (Nanda 2017).

Limitations of DHI

Though analysis of DHI anomalies appear simple, it can be at times hard to predict correctly the presence of hydrocarbon from seismic. Volcanic sills, streaks of calcareous sands, coal beds, over pressured sands and shales, thin-bed tuning and digenetic rocks can be all potential creators of bright and flat spots. The key validation factor is the polarity which is unfortunately often hard to determine correctly. An interpreter may have to consider all available seismic evidence for an integrated analysis to authenticate the DHI anomaly as a justifiable hydrocarbon prospect for drilling. It is also equally important to evaluate the prospect in the context of the overall geological perspective in the area to reduce exploration risk. The seismic validation and the geological play assessments must be mutually compatible before taking a drilling decision. A DHI anomaly, duly validated by geophysical

means, but without any geological likelihood of its association with a trap may have to be cautiously cast off. For instance, high amplitudes ('bright spot') and flat reflection events ('flat spot') seen at trough instead of the crest of a structure can be considered as suspects, unless a geological case can be made out for a unique type of stratigraphic trapping.

Even though the high amplitude 'bright spot' anomalies, are accepted as the most common yardstick for gas indication, they cannot be generalised. The high amplitude anomalies may only appear under certain favourable geological conditions, specific to an area. For an eligible classical 'bright spot' anomaly to materialize in seismic, a typical clastic depositional environment is a prerequisite where preferably the reservoir sand is highly porous and unconsolidated, capped by shale to create a significant negative impedance contrast. Mio-Pliocene and younger sands occurring at relatively shallow depths and exhibiting low impedances commonly qualify the best to meet the ideal geological conditions to create DHI anomalies. Nevertheless, different geological settings may have varied lithofacies and diverse impedance contrasts, which will eventually determine the nature of amplitude anomaly. In this context, it may be underscored that the rock properties of the shale overlying the reservoir (usually the crux of the study) needs to be closely investigated as it can have a wide range of lateral varying properties in different basins, capable of causing various amplitude patterns in DHI anomalies (Chapter "Shear Wave Seismic, AVO and Vp/Vs Analysis"). Shales, especially at shallow depths, are also known to exhibit significant anisotropy at times, which can further complicate the issue.

To offset the problem of false amplitude anomalies misleading as indicator of gas/oil, more sophisticated techniques such as, AVO (amplitude vs. offset), shear wave analysis and inversion techniques are developed to substantiate the prospects and reduce risk. These techniques are briefly discussed in Chapters "Shear Wave Seismic, AVO and Vp/Vs Analysis", "Analysing Seismic Attributes" and "Seismic Modelling and Inversion". Yet, another major shortcoming of DHI anomalies is the estimation of gas saturation in a reservoir from seismic response. A gas saturation of as low as $\sim 10\%$ tends to show similar responses as increasing saturations up to 100% (Chapter "Seismic Wave and Rock-Fluid Properties"). In general, the rock compressibility does not perceptibly change with higher gas saturation. Consequently many wells drilled based on amplitude anomalies though resulted in gas finds, were of too low in saturation to be commercially produced.

References

Anstey AN (1977) Seismic Interpretation, The physical aspects, record of short course "The new seismic interpreter". IHRDC, 4.1–4.24

Bulloch TE (1999) The investigation of fluid properties and seismic attributes for reservoir characterization. Thesis, Michigan Technological University

Clark VA (1992) The effect of oil under in-situ conditions on the seismic properties of rocks. Geophysics 57:894–901

Ebrom D (2004) The low-frequency gas shadow on seismic sections. Lead Edge 23:772

Gregory AR (1977) Aspects of rock physics from laboratory and log data that are important to seismic interpretation. AAPG Memoir 26:23–30

Liu Y (1998) Acoustic properties of reservoir fluids. Ph. D. Thesis, Stanford University

Nanda NC, Wason A (2013) Seismic rock physics of bright amplitude oil sands—a case study. CSEG Recorder 38(7):26–32

Nanda NC (2017) Quantitative analysis of seismic amplitudes for characterization of Pliocene hydrocarbon sands, Eastern Offshore, India. First Break 35 (9):39–45

Osif TL (1988) The effect of salt, gas, temperature, and pressure on the compressibility of water. SPE Reservoir Eng, 175–181

Rutherford SR, Williams RH (1988) Amplitude-versus-offset variations in gas sands. Geophysics 54:680–688

Schroot BM, Schüttenhelm RTE (2003) Expressions of shallow gas in the Netherlands North Sea, Netherlands. J Geosci 82:91–105

Strecker U, Knapp S, Smith M, Uden R, Carr M, Taylor G (2004) Reconnaissance of geological prospectivity and reservoir characterization using multiple seismic attributes on 3-D surveys: an example from hydrothermal dolomite, Devonian Slave Point Formation, northeast British Columbia, Canada, CSEG National Convention, pp 1–6

Wandler A, Evans B, Link C (2007) AVO as a fluid indicator: a physical modelling study. Geophysics 72: C 9–C 17

Wang Z (2001) Y2K Tutorial- Fundamentals of seismic rock physics. Geophysics 66:398–412

Whang LF, Lellis PJ (1988) Bright spots related to high GOR oil reservoir in Green Canyon. In: 58th SEG Annual International Meeting, Expanded Abs, pp 761–763

Borehole Seismic Techniques

Abstract

Borehole seismic deals with recording seismic waves using geophone(s) in a borehole with the energy source either at surface or in the borehole. When it is recorded with the energy source at the surface, it is known as well seismic survey and recorded with source in another nearby borehole, is known as cross-well survey. The well seismic surveys include check-shot and VSP surveys which measure the vital true overburden velocity, most essential for time to depth conversion and calibration of seismic reflections with well. It also allows to derive accurately the true interval velocities which are crucial to determine formation properties and also for effectual prestack depth migration for providing improved high resolution and high fidelity seismic images. The check-shot survey and the more accurate VSP surveys of different types with recording lay outs and processing involved are outlined. VSP survey applications in seismic calibration using corridor stacks and other advantages such as helping better data processing and predicting the geologic formations 'ahead of the drill bit' are described along with limitations.

The distinction between check-shot and VSP velocity surveys and the dissimilarities between well (seismic) and sonic velocities are brought out. The practice of constraining check-shot velocity function (T-D curve) with the lithologs and the derived interval velocities allows check for geologic consistency and improve the curve plot, particularly for thin formations is also discussed. Cross well surveys include tomography surveys and microseismic surveys.

Cross-well surveys using tomography technique for monitoring thermal stimulation for enhanced oil recovery (EOR) of heavy oil and microseismic techniques used for monitoring hydraulic fracturing of tight reservoirs for producing hydrocarbon are briefly stated.

Borehole seismic techniques deal with the recording of seismic waves using geophone(s) in a well and the energy source either at the surface or in the well. The former layout of source at surface and detector in the borehole is commonly known as *well seismic* survey and recording with source in one and detectors in another borehole is called cross-*well* survey, discussed briefly at the end of this chapter. Well seismic surveys measure directly the true overburden (average) velocity of seismic waves at a well and are of two kinds, the *check-shot* survey and the *vertical seismic profile* (VSP) survey.

N. C. Nanda, *Seismic Data Interpretation and Evaluation for Hydrocarbon Exploration and Production*, Advances in Oil and Gas Exploration & Production, https://doi.org/10.1007/978-3-030-75301-6_7

Check-Shot Survey

Acquisition Layouts and Energy Source

The check-shot survey, also known as well velocity shooting, is conducted to measure the true velocity which is a prime requirement for several applications (discussed later), the most important being conversion of seismic reflection times to depths of geologic formations. With a source firing at the surface, it records the first arrivals of direct waves (the first breaks) by geophone deployed in the well that is moved to different levels of depths. The geophone spacings are usually at irregular and at large intervals, restricted to the interested target horizons to save stand-by time of the drilling rig and linked cost. The surveys are mostly carried in vertical wells with zero and non-zero offsets though they can be conducted in deviated wells albeit with attendant operational problems. Energy sources deployed on land are usually dynamites or air-guns and in offshore it is the air-gun. While the

dynamite source is preferable due to its creating strong energy pulse, explosions can be detrimental for safety of the borehole. Lay-out of check-shot surveys with source-geophone configuration for vertical (zero and nonzero offsets) and deviated well are schematically illustrated in Fig. 1.

Ideally, on land the explosive energy source, placed as close as possible to the well-head of a vertical well (without risk to well), would cause a near-vertical ray path travelling down to geophones to measure the true vertical velocity (Fig. 1a). A source placed at an offset, to safeguard the borehole, would however, involve a slant travel path through the formations and transmission (refraction) effect, impeding measure of exact vertical velocity. A simple correction for the slant ray geometry can be made for computing the corresponding travel time for the vertical ray path from the knowns, the offset and the geophone depth in the well. However, it does not account for transmission effect of the ray path which is actually longer, and results in showing

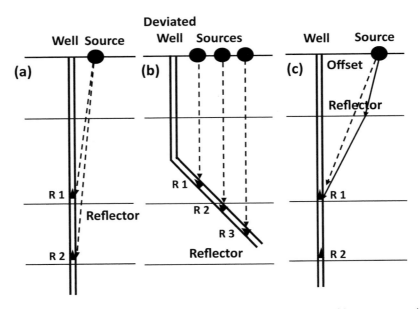

Fig. 1 Schematic showing Check-shot survey with source–geophone lay-outs. **a** Vertical well, Zero-offset source, **b** Deviated well, source position changing with each shot and **c** Vertical well, source with offset. R1, R2, R... show the receiver positions and dashed lines indicate the direct arrivals. Note in deviated well **b** the need to change source positions with respect to receiver after each shot to attain vertical ray path for measuring true average velocity. In **c** the direct arrival ray path is bent due to transmission effect and assuming it as straight would result in slower velocity. The effect gets more pronounced with increasing depth

slower velocity (Fig. 1c). The inaccuracy due to transmission effect to provide true vertical velocity can be critical as it gets more and more pronounced with increasing depth. Explosives, a powerful energy source while being an ideal choice may require large offset to avoid damage to the borehole which, however, may impact accuracy of measured velocity. It thus poses operational problem seeking for a trade-off between choices of energy source for on- land well surveys.

The dynamite source is blasted in (shot) holes drilled to depths below the weathering zones, (~ 18–30 m) and the wave travel time recorded by the receiver at a depth level, usually starting from the bottom of the borehole, called a shot. After a few shots, the shot hole collapses entailing many more holes to be drilled which is a further constraint for using explosive source and urging for alternative sources such as Vibroseis and air-gun sources. However, Vibroseis trucks are often not available and that leaves only the air-gun source which is most commonly used. Air-gun, the usual source in marine surveys, is usually a safe option despite its low energy strength and is deployed in a shallow pit, dug and filled with water. Air-gun source efficiency being poor and in areas particularly where the shallow subsurface formations show large attenuation, the generated source waves may not be able to produce good first-breaks for pick. In the case of deviated well, where the source to receiver travel path is slant, the source needs to be shifted for each receiver position so as to be vertically placed above which adds to operational problems of drilling multiple shot-holes that cost more time and money (Fig. 1b). This is one more reason for preferring air-guns to explosive as source.

The receivers are geophones, locked to borehole wall for better contact with formation and geophone placings are chosen preferably close to the major litho-boundaries so that accurate formation velocities can be derived. The receiver spacing is commonly irregular and random, typically 75–150 m (Brewer 2002), closer in interesting zones with relatively thinner layers. The number of shots planned depends on a trade-off between number of shots to be recorded for

more detailed measurement of velocity and the additional time and cost incurred due to drilling more shot holes and the ensuing rig- standby time for the extended operation. The shot starts at the deepest level, the bottom of the well, and continues upwards with the geophone moving to next shallower level after each shot is completed. Occasionally, repeat shots are taken at the same depth as quality check and the survey may typically comprise of around 20–30 shots in a well drilled to 3 km depth and if everything goes smoothly, the survey takes about 4–5 hours to be completed.

The T-D Curve (Velocity Function)

Direct travel times of seismic waves from the surface (source) to the geophones in the well, after due corrections for source-receiver geometry, weathering corrections (onland) and seismic reference datum (SRD), provide accurate average velocities at depths measured. A graph plotted with arrival time recorded at the geophone versus its depth, with reference to the seismic datum, known as the T-D curve, provides the average velocity function, the key to crucial conversion of seismic times to geologic depths. The seismic reference datum varies in onland data depending on terrain and general elevation of the area and is usually the mean sea level (MSL), though can be different depending on the areas. Since the data point intervals are large, joining the time values to obtain the velocity function may not be accurate as it may miss intermediate velocity breaks of some smaller formations (Fig. 2). Joining of the curve can be made more accurate to represent the smaller formations getting guided by the litholog plotted by its side. From the prepared T-D function, the computed interval velocities, viewed against formation lithologies, interestingly, can act as a check for the accuracy of the T-D graph drawn. The interval velocities without geologically meaningful values would reveal inexactness of the velocity function in that interval. To arrive at the best velocity function derived from the check-shot survey and with reasonable confidence, it is advisable to prepare

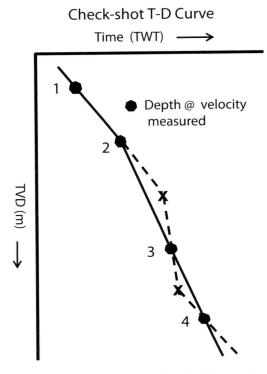

Check-shot T-D Curve

Time (TWT) ⟶

● Depth @ velocity measured

TVD (m)

1

2

3

4

Fig. 2 Figure illustrating T-D (time-depth) curve from check-shot survey for obtaining average velocity function. Because of large geophone spacing the line joining the recorded time values can provide imprecise velocities at intermediate levels. Note·the possibility of more segments as shown by dotted lines had the geophones were spaced closer (marked by X). Constrained with litholog and computed interval velocities for the segments could help check accuracy of the curve plotted and modify for better details

composite panel displaying the T-D curve, the litho-column at its side and the interval velocity, overlain. However, if a sonic log is available it can also act as guide and check for the T-D curve and may be displayed in the composite panel to build confidence. The interpreter may be mindful of this simple but important practice because often the data is imported to work station digitally and directly for depth conversion and other uses without scope for checking and supervising the velocity function.

Benefits of Check-Shot Surveys

The check-shot data, besides being essential for time-depth conversion and evaluation of formation properties by derived interval velocities, it is also plays significant roles in other related applications. It provides the time from the surface for integration of sonic log times which are usually not logged from the surface datum but from depth of around 1000–1200 m. The T-D curve is also used for correcting the sonic velocities (described later in the chapter), which are used for preparation of synthetic seismograms and continuous velocity logs (CVL) for well-seismic tie. Instantaneous velocity (CVL) logs duly constrained by check-shot data helps transform it from depth to time domain enabling accurate estimation of interval velocities for the corresponding thin layers in seismic, otherwise not workable from check-shot survey due to recording of data at large spaced interval.

Vertical Seismic Profiling (VSP)

Vertical seismic profiling is similar to a check-shot survey but is a more evolved and advanced technique meant to offer solutions to diverse problems in hydrocarbon exploration and production. The VSP, unlike in check- shot, records all seismic arrivals, the direct, the reflected and the multiples by the geophones in a well, placed in a vertical array of short regular interval. Figure 3 is a schematic diagram illustrating the VSP zero offset survey layout in a vertical well with source and geophone configuration. However, several other types of VSP survey layout geometries can be designed for use in both vertical as well as in deviated and horizontal wells depending on the exploration objective. Zero offset VSPs conducted in vertical wells with small offset are more common and is described in detail while a few of the other type of VSPs are briefly outlined.

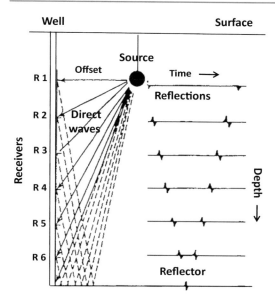

Fig. 3 Diagram of VSP survey layout with source-geophone configuration illustrating the recording process. Geophones (R1, R2,…) are placed in the well with close and regular spacing. The solid lines stand for downgoing waves, the direct arrivals (first breaks) and the dashed lines for upgoing reflections from a horizon. Note the reflection from the bed close and below the receiver is a one-way propagation (R6) unlike the two-way propagation in surface seismic (Modified after Balch et al. 1981)

Zero Offset VSP (ZVSP)

The zero offset VSP (ZVSP) acquisition is the simplest to design and commonly serves the exploration objectives, which amidst others, to provide mainly accurate true vertical velocity and the means to calibrate seismic with well. While check-shot records first arrivals at receivers, placed arbitrarily with relatively large and irregular spacing, the VSP survey, in contrast, samples all seismic wave arrivals at receivers which are placed regularly with close spacing usually ~ 15–30 m apart, (Brewer 2002). The geophones are pre- planned to be placed near interesting horizons and at regular and short spacing in the well for improved resolution. This is similar to surface seismic layouts, the difference being instead of sampling in horizontal direction the geophones sample in vertical direction in depth to provide resolution and accordingly named, vertical seismic profiling

(VSP). The VSP survey geometry allows the recording and processing of not only the downgoing direct arrivals but more importantly, the later arrivals of upgoing reflections from the rock boundaries present below the recording geophone depth (Fig. 3). The downgoing direct waves recorded at the geophone enable measuring the true average velocity and also records the source wavelet which offers valuable information about the varying shape of the source wavelet with depth. The source wavelet changes progressively with depth during propagation due to attenuation and the source signature recorded at different depths signifies the attenuation factor in the media. Analysis of changes in the shape of the downgoing source wavelet captured at each depth point, in addition to attenuation information, also leads to understanding anisotropy present in the subsurface at different levels. The plethora of information offered by the VSP can be used effectively and gainfully to optimise crucial parameters in seismic processing of data such as in deconvolution to obtain better resolution.

Source Wavelet Recording and Benefits for Seismic Data Processing

Some benefits offered by VSP for optimising seismic processing can be summed up as:

- Designing wavelet for deterministic deconvolution from known source wavelet,
- Identifying multiples and their genesis for effective removal,
- Determining attenuation spectrum of subsurface for 'Q' compensation,
- Processing for zero-phase reflectivity,
- Estimating accurate interval velocities, and
- Determining optimal filter bands for signal enhancement.

Of all the upgoing reflections recorded in VSP arriving from below the geophone at a particular depth, the reflection from the boundary close and below the geophone is recorded with only one-way propagation in the medium (refer Fig. 3). This is in contrast to conventional seismic recording which involves two-way propagation of the seismic wave. The one-way wave propagation entails reduction in absorption and transmission losses in the medium, avoids near-surface complexities and intra-bed multiples, and results in VSP offering better resolution than seismic.

Corridor Stack

As in surface seismic data, in a zero-offset VSP, along with the primary reflections, direct waves and multiples are also recorded. The downgoing and the upgoing waves are near symmetrical and with opposite dips (Fig. 4). The VSP data are meticulously processed separating the downgoing and upgoing waves, which are then utilized separately. The downgoing waves are the direct arrivals that yield the time-depth curve which is used for depth conversions and calibration of sonic log and computation of interval velocities

One way Time (sec) ⟶

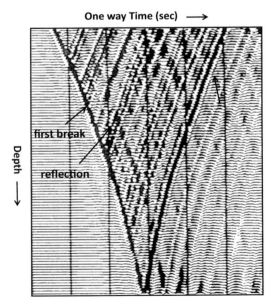

Fig. 4 Displays a VSP raw record showing downgoing (first break) and upgoing waves (reflection) in a VSP record. Note the symmetry of the downgoing and upgoing waves (Courtesy, ONGC, India)

similar to check-shot but more precisely because of closer grid data.

The upgoing waves after further processing finally display the deconvolved primary reflection events, time-shifted to two-way arrival time against depth (Fig. 5a), where the zero-phased reflection events are aligned in a horizontal direction. The source wavelets recorded at depths by the downgoing waves enable to make the reflection events zero-phase by applying deterministic deconvolution technique in data processing. The panel display is important as it signifies the quality of the VSP upgoing data, the interpreter is given to analyze. A corridor, usually of 200–300 ms time window is arbitrarily selected beginning at the first breaks along the outer edge of the upgoing VSP panel, and the traces falling within it are stacked into a single trace. This evades the multiples and is called the outer corridor or more commonly '*corridor stack*'. At the opposite end of the aligned upgoing data, a corridor can also be chosen to stack selected traces within the corridor and is referred to as the inner corridor stack that emphasizes multiples including the interbed multiples (Campbell et al. 2014).

It may be noted that the quality of reflection amplitude displayed in the stacked trace is influenced by the choice of the corridor width which could be subjective. It is desirable that the processed upgoing primary reflection panel along with the corridor used for stack be displayed to enable the interpreter to view the quality of VSP data in general and the corridor stack in particular, used for seismic calibration. Often Irregular corridors are chosen arbitrarily, called 'surgical window' to influence the amplitude of the corridor stack which may not be true representation of reflection amplitudes. A number of the stacked trace is repeated to form a segment for convenient display for match with seismic (Fig. 5b).

A corridor stack is similar to a synthetic seismic trace generated from well data and is multiple-free, zero phase 1D seismic response of earth's reflectivity and is considered the preferred tool for reliable calibration of surface seismic with well. However, though both surface and VSP belong to the same seismic system, arrival time of the

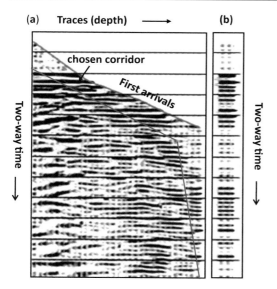

Fig. 5 Displaying VSP processed final panel **a** upgoing events and the downgoing first arrivals and **b** the (outer) corridor stack which is a repeat of the summed trace in the chosen corridor of 200–300 ms of time width starting from the first breaks. Note the influence of choice of corridor on the amplitude of the stack displayed in the corridor stack (**b**) which is used for seismic calibration. The remaining part of the upgoing wave panel known as inner corridor and emphasizes multiples and is important as it displays the VSP data quality (Image courtesy of ONGC, India)

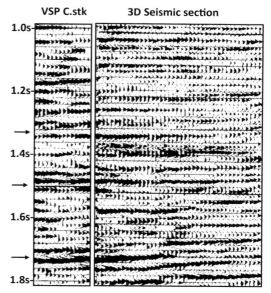

Fig. 6 Showing VSP corridor stack and seismic calibration. Note the wiggle with variable area mode of display of VSP and seismic stack which enables to perceive the quality of phase match (arrow mark **a**) with the 3D seismic reflection horizons

stratigraphic layers seen in the VSP recorded in the well may differ from the surface seismic time due to several reasons. These may include difference in response of the recording instruments and the detectors, disparity in processing related issues, improper statics and datum corrections and the ambient noise present in the two data sets. The corridor stack in such case, may require appropriate time-shift for reconciliation with seismic data. A corridor stack with good phase match with 3D seismic (peak and trough), in wiggle and variable area display mode is shown in Fig. 6.

Prediction ahead of Drill Bit

The most unique advantage of VSP is perhaps having the capability of 'predicting ahead of drill bit' as an aid to drilling engineers and exploration managers. Since reflections from all

discontinuities present in the subsurface are captured, a VSP can help predict the geology of formations below, yet to be drilled, within and beyond the planned target depth. This is achieved by the technique known as 'seismic inversion', a technique to transform seismic reflection amplitude to velocity (Chapter "Analyzing Seismic Attributes"). In some situations, where drilling further to deeper depth is contentious, considered futile or dangerous, it may be prudent to conduct a VSP survey to help decide the appropriate exploration strategy including drilling. One such instance can be running into an unexpected high-pressured zone during drilling before the target depth is reached. To continue further drilling blindly without knowing the degree and extent of the high-pressured zone could be a difficult and risky decision to take. Yet another case in point may be where the geologic object is not met even after drilling the planned target depth and the geoscientist insists that the objective may be a little deeper worth drilling. Such issues may require the management to take a decision to either close down drilling operations or continue

VSP
processed section

Cor. CVL
stk (m/s)

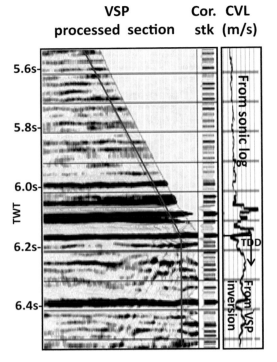

Fig. 7 Figure illustrating VSP prediction 'ahead of drill bit' in a deep offshore well. The Sonic was logged to drilled depth (TDD), beyond which no subsurface geologic information was available. Velocities extracted from inversion of VSP corridor stack amplitude is used for predicting rock properties of formations below, yet to be drilled. Sonic log taken later corroborates the prediction (Image courtesy: ONGC, India)

further. Abandoning the well without meeting the exploratory objective may be tremendously frustrating considering the time and money already expended. On the other hand, if a decision to continue drilling is to be taken it would require that all the geologic information about the subsurface formations is estimated properly to plan afresh a suitable and safe drilling strategy with revised target depth.

VSP survey provides the solution by using seismic inversion technique that quantitatively estimates interval velocities from the VSP recorded reflections, similar to a CVL from sonic log (Fig. 7). Constrained with the sonic curve, it can predict subsurface geology- formation lithology, thickness, porosity and more importantly the pressure ahead of the drill bit and help engineers plan suitably for fast and safe drilling.

Offset (Non-zero) VSP

The most commonly deployed survey geometries in a vertical well are the *zero- offset* but non-*zero offset* and *walk-away* VSPs can be designed depending on the exploration objective at hand. The zero offset VSP, as mentioned earlier, is recorded with the source placed close to well-head, similar to check-shot survey and provides a 1D seismic response at the well liken to a synthetic seismogram. A non-zero offset called Offset VSP, on the other hand, is recorded with an energy source placed several hundreds of meters away from the well-head and provides single fold 2D seismic section, though of limited areal extent close to the well. A schematic of non-zero offset VSP is shown in Fig. 8. Properly acquired and processed a corridor stack is relatively noise-free, high resolution, zero-phase data which makes the technique extremely useful in reservoir delineation and characterisation where conventional seismic resolution fails. Nonetheless, the lateral extent of the subsurface imaged is small, limited to less than half the source- well offset, and depends on factors like structural dip, source

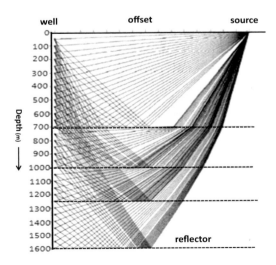

Fig. 8 Schematic showing ray tracing for an offset VSP model configuration showing the downgoing and the upgoing waves from an offset source and recorded by the detectors. Notice, the small areal extent of the subsurface coverage by the offset VSP which becomes smaller with increasing depth

offset, the depth of the reflector and the depth interval sampled based on placement of the shallowest and the deepest geophones in the well.

Walkaway VSP

Yet another type, the walk-away VSP, is a more elaborate technique in which seismic signals are recorded at each geophone in the well with a number of shots fired at the surface with varying offsets and along one or several azimuths around the well. The walk away survey geometry enables to generate stacked seismic section similar to conventional CDP and provides high resolution multi-trace and multi-azimuth coverage around the bore-hole for reservoir characterisation. The survey to be effective, needs careful configuration of source-receiver geometry but the massive increase in number of shots to be recorded. However, the survey can be a serious impairment to straining operational resources, time and cost.

Benefits of VSP

Carefully planned and properly executed, VSP survey can be highly useful in solving exploration and production problems. The VSP has several applications and the important benefits include -

(i) Obtaining vertical velocity function for precise depth conversion,

(ii) Optimising processing parameters for seismic data enhancement,

(iii) Calibrating seismic with well data,

(iv) Predicting formation pressure and rock properties 'ahead of drill bit',

(v) Delineation and characterization of reservoirs and

(vi) Providing surface time to integrate sonic velocity and drift correction.

The first four benefits are discussed earlier. However, its application for delineating lateral extension of a pay beyond the well bore, may be debatable because of its limited subsurface coverage. Imaging a mere 500 m extent of a reservoir requires a wide offset of the source of about one kilometre at well head offset which can be a large technical and operational constrain. Hydrocarbon reserves are swept usually from about 400–500 m around the well bore and conducting a VSP, under the circumstances, may not be much helpful unless the problem is geologically unique to be solved. However, there may be fluid flow problems during production and VSP may be helpful in improving the reservoir characterization by identifying minor faults, thin shale layers, highly permeable sand beds and fractures which cause the snags.

Surface time integration of sonic velocity is another area of application of well velocity. Sonic logs are seldom recorded beginning from

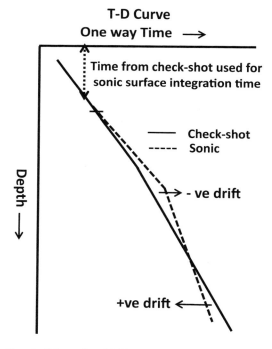

Fig. 9 Schematic of T-D curves of velocity function illustrating the utility of check-shot/VSP survey. Besides providing the time from the surface to integrate sonic time from ground reference it calibrates sonic by applying suitable drift corrections. Drift is the difference between check-shot and sonic times which can be nonlinearly variable. The drift correction applied to sonic is positive if sonic time is slower and negative if higher with respect to true times measured by the check-shot/VSP

the surface, and the sonic integrated time for the depth intervals would need the initial time from surface to be added for arriving at the velocity function from the datum. This is called the 'surface integration time' and the VSP/check-shot survey supplies this critical missing time for the interval for which the sonic is not logged (Fig. 9). This helps sonic velocity to be used for computing impedance series to prepare synthetic seismograms.

Check-Shot and VSP Survey Comparison

Though both the check-shot and VSP are well surveys that measures velocity in a well, and have similar operations, there are several notable differences between the two surveys and are summarized in Table 1.

Well (Seismic) Velocity and Sonic Velocity Dissimilarities

Generally, the well velocity measured at a well is found different from the sonic-derived velocity and this could be due to various reasons. Primarily, the seismic and sonic waves greatly vary in their source frequencies. The seismic uses a frequency range of ~ 5–125 Hz, which is much lower than that used in sonic of ~ 2–20 kHz, (Bulant and Luděk 2008). Since higher frequencies exhibit greater velocities due to dispersion effect, sonic is likely to indicate higher velocities than seismic. Secondly, and more importantly, the propagation principles and depth of investigations are highly dissimilar. The volume of rocks through which seismic waves travel, are much larger and may have diverse facies and propagation characteristics than the small volume of rocks through which sonic waves travel

Table 1 Check shot and VSP surveys—comparison

S. no.	Check-shot	VSP
	Source close to wellbore	Source placement variable depending on objectives of survey
2	Geophone spacing large (~ 50–100 m) and irregular	Geophone spacing close (~ 10–20 m) & regular
3	Records direct arrivals (first breaks)	Records direct and later arrivals (subsurface reflections)
4	Source signal not recorded	Source signal recorded and its variation with depth provides measure of absorption (Q) in the formations
5	1D first break recording; no information on subsurface	1D/2D/3D recording of reflections, subsurface imaged close to well in 2D/3D
6	Processing trivial- picking first breaks and static-corrected to datum	Elaborate processing of down-going and up-coming waves with static corrections, decon, NMO etc., as in conventional seismic
7	Measures average velocity at depths in a well and delivers T-D curve	Measures velocity more accurately, delivers T-D curve and a seismic section or volume depending on 2D/3D type of VSP
8	Used for tying well events to seismic time, sonic drift corrections and sonic surface integration time	Used for all the purposes as in check-shot plus more precise well-seismic calibration, optimizing seismic processing, reservoir delineation and characterization and prediction ahead of drill bit
9	Survey simple—entails less time and money	Relatively more involved—takes more time and is expensive

(Thomas 1978). Seismic velocities are thus average velocities over the distance the seismic energy propagates through the different formations, whereas the sonic measures accurately the interval velocity of the formation it travels immediately close to the borehole. Phenomena like absorption and short-path multiples, involved in seismic wave propagation, but not encountered in sonic measurement, also tend to make seismic velocity usually lower. Furthermore, fractures (open) present in a medium, are also likely to lower seismic velocity, while the sonic velocity remains unaffected because the sonic wave travels the path of least time which is through the matrix, avoiding the fractures. Incidentally, this attribute of sonic, insensitive to fractures, is used to calculate fracture porosity, computed by the difference between the neutron logs derived total porosity and the sonic derived intergranular porosity.

Sonic Drift Correction

Though the sonic velocities can be expected in general to be faster than seismic, for the reasons cited above, sometimes the sonic velocities can be slower. The sonic velocity can be slower in cases of poor well conditions with mud-invaded and altered zones, though the problems are mitigated to a large extent by present day, better developed sonic tools. Sonic velocities can also be affected by errors due to tool sticking, washed-out zones and cycle-skips (Brewer 2002). Cycle-skipping in a sonic log occurs when the first arrivals at one or both the receivers are too weak to be detected (skipped). Sonic measures the slowness (μs/ft, the time interval) between the two receivers on an array and depending on the skip of the wave arrival instance at the first or at the second receiver, the velocity may be faster or slower.

As mentioned seismic and sonic velocities can be different and the differences observed between the well seismic and the surface integrated sonic velocity functions, represented by the time-depth plots, the T-D curves, is called the 'drift'. The sonic drift from the seismic velocity function

needs correction which is dynamic, varying in signage and magnitude over the range of well depths. It can be positive in some intervals and negative at other levels at the well (Fig. 9). This necessitates proper corrections to sonic velocities prior to its use for computing the impedances to generate the reflectivity series for preparing synthetic seismograms (Chapter "Seismic Interpretation Methods") for seismic-well calibration.

The prime difference between the seismic and sonic velocities may be restated that they measure fundamentally two different kinds of velocities. Well velocity (VSP/check shot) survey measures the true vertical (average) velocity in rocks in the vicinity of the well whereas the sonic continuously measures the variations in interval velocity of formations (CVL), limited to rocks immediately adjacent to the borehole. The differences between well velocity and sonic velocity are tabulated in Table 2.

Limitations of Check-Shot and VSP Surveys

The velocities measured in a well, although generally reliable, may however suffer from inaccuracies due to poor quality recording, picking signals and processing of the first breaks and later arrivals. The recording quality can be affected by bad geophone coupling, inadequate source strength, adverse borehole conditions and ambient noises in and around the well. Often the source strength may be insufficient to impart enough energy into the subsurface at greater depths for propagation, particularly onland, but unfortunately it cannot be helped as enhancing source strength may jeopardize the safety of the well. Especially in onshore wells, the seismic velocity measurement may be contaminated due to the locally present near-surface anomalous velocity zones, shallow anisotropic strata, improper datum corrections, and propagation delays. Velocity dispersions and short-path multiples can also significantly delay seismic arrival times. As a reliability check, it may be sensible for the interpreter to use the well

Table 2 Well velocity and sonic velocity dissimilarities

S. no.	Well seismic	Sonic
1	Source: Low frequency ∼5–125 Hz	High frequency, ∼2–20 KHz, causes dispersion
2	Depth of investigation: Large volume of rocks	Small volume of rock in the near vicinity of the well bore
3	Propagation effect: Suffers from attenuation and short-path multiples during long travel	No such effect for the sonic wave, short travelling in the formations
4	Propagation mechanism: Body wave travelling in the media and affected by fractures	Head wave which takes the least time by travelling through the matrix and not affected by fractures
5	Anisotropy: Seismic velocity affected by anisotropy	No anisotropic effect on sonic velocity
6	Velocity: Measures average vertical velocity over the distance travelled through different formations	Measures Interval velocity (slowness) continuously for formations through which it travels
7	Deliverables: Average velocities at depths versus time; T-D curve and average interval velocities	Log of continuous recording of slowness in formations, instantaneous velocity log (CVL) and interval velocities
8	Corrections: recorded from the surface and needs appropriate static corrections involving weathering zone and terrain	Needs the time from well velocity to be integrated for providing T-D curve from surface and drift corrections due to cycle skipping
9	Utility: Conversion of time to depth, interval velocities, providing drift corrections and surface integrated time to sonic data	Computation of earth reflectivity series to construct synthetic seismograms for seismic calibration and interval velocities

velocity by checking ties at all the levels of seismic markers with the geologic boundaries at the well rather than just the one at the target level. The formation velocity computed from average velocity, particularly from check-shot T-D curve, as stated earlier, can be a good check for velocity data reliability. However, the work stations usually import the velocity data digitally and the option of reviewing the data quality may be missed if the interpreter is not watchful.

Before attempting a customary match of the VSP corridor stack with seismic for tie, it is important the interpreter take a look at the processed final upgoing wave section and not the corridor stack alone. This provides an insight into the quality of recorded and processed data for judging the reliability of the reflection events in the VSP section vis-a-vis the chosen corridor. As stated earlier, as the corridors are chosen arbitrarily, the processor's bias influences the corridor stack, particularly in the deeper sections below the drilled depth where reflection quality deteriorates. The final processed upgoing wave panel works outs as a quality check (QC) for the VSP and the chosen corridor stack which is crucial for seismic calibration. It can be best

achieved by a composite panel display including the VSP processed upgoing wave section with first breaks, the corridor stack and the seismic section to rationalize the VSP calibration with seismic (Fig. 10). The mode of display of seismic and VSP corridor stack sections is also important. It is important to use wiggle and variable area display rather than the variable density to validate proper character correlation of object reflections.

VSP surveys, although are likely to offer more reliable and accurate information, seem to be under-utilised in practice for reservoir characterization. In addition to earlier stated operational and economic constraints in the relatively more time-taking VSP acquisition, the other reason may be lack of definition of the exact survey objective and inadequate planning. Nevertheless, a properly executed survey for achieving effective solutions may require a lot of rig down-time which can be prohibitively expensive and therefore often skipped. Many companies seem to be content with recording a relatively simple and cheaper check-shot survey or a VSP with moderate sampling rate with the sole aim to obtain the velocity, as a cost- benefit trade-off.

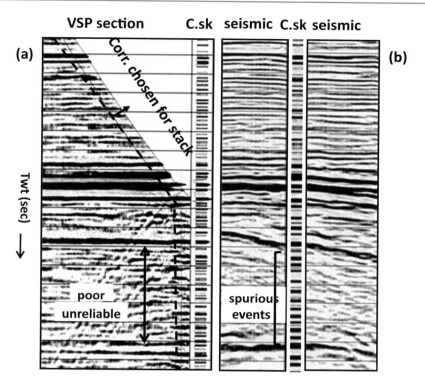

Fig. 10 Composite display of VSP, corridor stack and seismic for quality check. Note the noisy and unreliable data recorded in the lower part (**a**) that show a plethora of strong events in Corridor stack (**b**). Absence of these events in seismic corroborates the corridor stack events as spurious (**b**). Notice the effect of choosing corridor on the stack (**a**) which can be highly subjective and can generate spurious events (**b**). Also note the difficulty in matching phase of events due to the mode of display in variable density and also the slight VSP-seismic time mismatch needing correction of VSP time (Image courtesy of ONGC, India)

Cross-Well (Borehole) Survey

The cross-borehole or cross-well seismic survey involves two boreholes and is carried by placing a seismic source in one well and a string of receivers in a nearby well to record direct arrivals and reflections (Fig. 11). Innovative and improved acquisition and processing systems have made it possible to measure accurate travel times with non-destructive, high frequency (kHz) sources and sensitive geophones, both lowered inside the wells. Tuned air guns and piezoelectric elements are commonly used as sources with wide-band geophone arrays as receiver. The receiver array is held fixed in one well while the source in the other well is changed by moving upwards, starting from the bottom

similar to conducting check-shot/VSP, and the shot incited each time for recording data. After completing one round of shooting, the receiver array is relocated and the source incitement round is repeated. The source spacing is small, typically between 2.5 and 20 ft with similar receiver spacing (Jerry and Langan 2001). The source and receivers each, thus, occupying a multiple number of shots, spaced over a depth range of the zone of interest, generate a huge amount of data which is processed similar to surface seismic.

Since both the source and receiver are placed inside hole, the cross well data does not suffer the distressing problems due to the near and shallow surface effects, as in VSP (one way) and surface seismic (both ways). Additionally, the short separation between the wells, offers much higher

Fig. 11 Schematic diagram showing cross-well seismic reflection tomography survey lay out of source receiver configuration in the target window. For each shot, the direct and reflected waves are recorded at the geophone array and after all the shots are incited, the geophone array is changed and the process of firing the shots and recording at geophones is repeated. This generates huge amount of data for tomographic mapping of interval velocities accurately

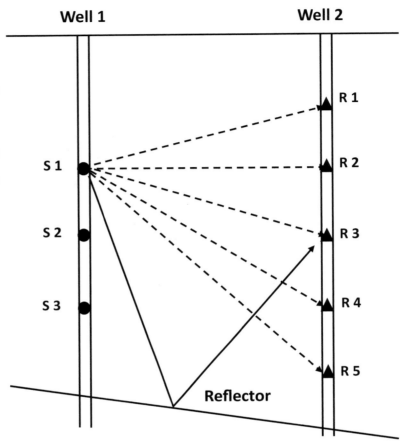

vertical resolution than the VSP data. The spacing of the shot and receiver together with the vertical interval they cover (aperture) provides the spatial resolution. Essentially, the cross well data yields accurate velocity field and precise seismic images of geologic objects through a technique known as tomography. Depending on the objective, it can be traveltime tomography or reflection time tomography.

Traveltime Tomography

The technique uses the recorded direct travel times between the source and the detector and reconstructs a velocity model through travel time inversion. As is customary in inversion processes, an initial velocity model is presumed and the recorded travel time data are compared.

The initial model is iteratively modified until the best possible match is achieved. The two-dimensional velocity field, so computed between the wellbores, is represented usually by a map known as a tomogram.

Reflection Tomography

Reflection tomography, also known as reflection traveltime tomography (RTT), on the other hand, involves reflections from the horizons below which deals with two-way propagation unlike the one way propagation in traveltime tomography. Similar to traveltime tomography an initial earth model is built and computed traveltimes of the reflections are matched to the picked traveltimes and reflector positions shifted iteratively till good match is found. This results in providing the interval velocities for the layers which ultimately help in prestack depth migration of seismic data for producing better images. Reflection tomography requires a more intense data processing and offers better accuracy.

Tomography Applications

Travel time tomography provides velocity map of the intervening zone between the two wells which is useful in preparing detailed and accurate geological cross sections between the two wells for reservoir monitoring and solving production problems. One of the important application is in monitoring enhanced oil recovery (EOR) processes for their sweep efficiency, particularly in production of low viscosity heavy oil from tight reservoirs (see Chapter "Oil sands, Heavy oil, Tight Oil/gas, BCGAs and CBM"). Production of heavy oil necessitates injection of steam in a well to mobilize oil and its efficiency depends on the proper progress of the heat front and sweep. However, the monitoring needs two tomography surveys, a base-line survey for estimating initial reservoir properties and a repeat (time-lapse) tomography after a period of time to evaluate the changes induced due to the thermal stimulation. The principle is somewhat similar to time lapse 3D surface seismic (4D) but with the major contrast, the sources and receivers are located in the borehole. Consequently, cross-well seismic does not suffer from the limitation of the requisite repeatability of 3D acquisition parameters in surface seismic and delivers more reliable estimate of changed mechanical rock properties. Differences in travel times as small as 0.5 ms in the lapse cross-well survey, attributed to the effect of steam flooding could be noticed (Kiyashchenko et al. 2013).

Reflection tomography provides a more robust and accurate velocity field which results in superior imaging of subsurface geology. Important applications include reservoir characterization and field production by analyzing the heterogeneity of the reservoir, fractures and linked anisotropy that controls permeability and ultimately the fluid flow patterns. Anomalous high-permeable and impermeable (barrier) paths that can lead to premature water-cuts that impedes production can be detected and the development plans suitably modified. Another significant application of reflection tomography technique, besides cross well surveys is in the seismic data processing of prestack depth migration (PSDM), described in Chapter "Evaluation of High-Resolution 3D and 4D Seismic Data", in which reliable interval velocity is the mainstay of efficient migration for creating accurate seismic images and depths.

Microseismic Surveys

Microseisms in the context of hydrocarbon exploration refers to rock-deformation caused by induced stress likened to natural earthquakes of very small magnitude. The microseismic method involves passive monitoring of microseismic events caused by reservoir stimulation processes such as hydraulic fracturing used for producing hydrocarbons from tight reservoirs. During the stimulation process the pressure is applied externally to the reservoir to create small scale disturbances, called the microseisms, which result in deformation, including fractures and are detected by very sensitive geophones. The method is also known as *passive seismic monitoring*.

The survey deploys an array of 3-component geophones typically of low frequency, at 2–10 Hz and very high frequencies (hundreds of HZs), broad band flat responses depending on whether they are planted inside borehole or on surfaces. The survey records the P and the S-waves, the latter being more sensitive for mapping fracture geometry and recording is done fast with very high sampling rate. In a way it can be likened to 3C seismic survey being conducted underground across wells. In borehole planting, the geophones are deployed near the zone of interest in the observation well which is adjacent to the well where hydraulic fracturing process is carried out. Receivers placed in borehole are sometimes permanently installed in the well due to the long periods of time needed for such surveys. Surface deployment of geophones, on the other hand, is generally large in numbers and has the innate advantage of extensive azimuthal coverage for better resolution and detecting anisotropy. While placement of geophones in boreholes has the advantage of better signal-to-noise characteristics due to the closer proximity to the microseismic sources, the surface recording is significantly more cost effective as there is no need to drill observation wells (van der Baan et al. 2013).

Monitoring 'Fracking'

Microseismic surveys are used extensively to monitor reservoir stimulation methods such as hydraulic fracture, known as 'fracking', of tight reservoirs, especially unconventional shale reservoirs for producing hydrocarbon (Chapter "Shale oil and gas, Oil shale and Gas Hydrates"). The data are recorded at high sampling rates by the geophones that detect the P and S waves generated during stimulation process of hydraulic fracturing which induces stress (St-Onge and Eaton 2011). Microseismic monitoring is essentially to know the location of the microseismic events (the fractures) and their magnitude which is a measure of size of the events. Magnitude is measured in terms of how much energy is released and is determined from the P and S wave amplitudes to interpret the fracture dimensions. The location which signifies the extent and orientation of the fractures is decided by the traveltime of the waves. However, for this to achieve a velocity field is needed and creating an accurate velocity model is a primary requisite. This is achieved by constraining the velocity model with sonic and surface seismic data.

Limitations

Hydraulic fracturing is a complex reservoir stimulating process involving geomechanical

rock properties, pore pressure and insitu stress. The tensile fractures, their dimensions and orientations can be a challenge for prediction of the fracture geometry created in the rock. The hydraulic fracture is predominantly a high energy and relatively slow process (long hours), where microseismic is relatively a fast recording process (milliseconds) with relative low energy. Further, the hydraulic fracture is dominantly tensile opening whereas the microseismic is often a shear dominated mechanism. According to Maxwell (2011), therefore there can be a number of paradoxes between the rationalization of hydraulic fracture and modes of microseismic deformation.

References

Balch AH, Lee MW, Miller JJ, Ryder RT (1981) Seismic amplitude anomalies associated with thick First Leo sandstone lenses, eastern Powder River Basin, Wyoming. Geophysics 46:1519–1527

Brewer JR (2002) VSP data in comparision to check shot survey, search & discovery article, # 40059, 1–5

Bulant P, Luděk K (2008) Comparison of VSP and sonic-log data in nonvertical wells in a heterogeneous structure. Geophysics 73:19–25

Campbell A, Leaney S, Gulati J, Podgornova O, Leslie-Panek J, Von Lunen E (2014) 3D VSP in an unconventional setting: images, anisotropy, Q, multiples and full waveform inversion. CSEG Recorder 39

Jerry HM, Langan RT (2001) Crosswell seismic profiling: principle to applications. Search Discov. Article #40030.6

Kiyashchenko D, Lopez J, Adawi R, Rocco G, Gulati J, Campbell A (2013) Borehole time-lapse seismic tracks steam flood. Hart Energy

Maxwell Shawn C (2011) What does microseismic tell us about hydraulic fracture deformation. CSEG Recorder, View issue, 36

Stewart RR, DiSiena JP (1989) The values of VSP in interpretation. Lead Edge 8:16–23

St-Onge A, Eaton DW (2011) Noise examples from two microseismic datasets. CSEG Recorder, View issue, 36

Thomas DH (1978) Seismic applications of sonic logs. Log Anal 19:23–32

van der Bann M, Eaton D, Dusseault M (2013) Microseismic monitoring developments in hydraulic fracture stimulation, INTECH, Chap. 21, pp 439–466. https://doi.org/10.5772/56444

Reservoir and Production Seismic

Evaluation of High-Resolution 3D and 4D Seismic Data

Abstract

3D seismic data is acquired with regular, close grid spatial sampling which allows application of high-end processing techniques to deliver high resolution high density data. The superior seismic images help reservoir characterization by quantitative estimate of reservoir rock-fluid properties which are crucial inputs to reserve estimates, reservoir modelling and simulation for planning suitable production profile. The 3D seismic is often synonymously called 'Reservoir Seismic' or 'Reservoir Geophysics'.

3D seismic is a volume data which can be interpreted three dimensionally. It can also display 2D seismic sections made out from the volume data in any desired azimuth for vertical viewing. Horizontal viewing of data of slices cut out from the 3D cube is another major advantage which resolves small-scale depositional features better in plain view than in 2D vertical sections. Such stratal attribute slices are extensively used to map channel/fan complexes with their associated diverse facies. Horizontal-view seismic is also useful for sequence stratigraphy interpretation (SSSI) to build tectono-stratigraphic frameworks for petroleum system modelling. High resolution and density of 3D seismic data allow extraction of multiple seismic attributes that can predict reservoir rock-fluid properties more precisely. One of the prime reason for much

enhanced 3D seismic images is migration of prestack data in depth domain (PSDM) and is outlined with shortcomings. 3D volume interpretation and visualization of data is holistic and in contrast to profile-wise interpretation of 2D seismic, is fast, more accurate and reliable. Work-flow for 3D data interpretation and its applications and advantages are deliberated with seismic illustrations and examples.

4D seismic is a time-lapse repeat 3D survey which evaluates the changes in the reservoir properties (dynamic characterization), under production. The 4D seismic is helpful in studying reservoir drive mechanisms during production and their imminent impact on fluid flow patterns, known as seismic reservoir monitoring (SRM). During production reservoir parameters undergo changes (Reservoir Geomechanics) that impact fluid flow pattern and 4D can estimate the changed parameters required for reservoir dynamic characterization. 4D seismic is therefore synonymously called 'Production Seismic' or 'Production Geophysics'. 4D has also other applications such as identifying areas of by-pass oil and flow barriers and monitoring EOR sweep efficiencies. These are dealt in the chapter along with limitations of 4D seismic monitoring.

High resolution, high-density 3D and 4D data, which is a time lapse 3D, offer scopes for precise estimation of rock and fluid parameters, crucial for reservoir characterization and monitoring during development and production stages of field. The 3D surveys are relatively expensive and are generally carried out post 2D, in limited prioritized areas for defining in detail the exploration targets of interest. This includes prospects which are found hydrocarbon bearing as well as the potent ones likely to strike hydrocarbon. However, with continuing innovative and improved techniques, the 3D acquisition and processing costs are considerably reduced and often the 3D data are acquired in larger areas in initial stage of exploration itself. This has also an advantage that there is no need to go back to the area for reacquisition in case of discoveries and avoid associated environmental and administrative hassles.

The basic interpretation and evaluation workflow remains essentially similar to those of 2D seismic except for it is an interactive interpretation process with workstation using more sophisticated techniques. This requires a multidisciplinary synergetic approach, involving all types of data such as seismic, geological and geochemical, well log, core, drilling, reservoir and production engineering data. Obviously, the evaluator is required to have knowledge, expertise and experience, and more importantly, an attitude to work in a team of persons from diverse disciplines, to produce the desired results. Since 3D data is increasingly used for reservoir characterization and 4D for monitoring fluid flow during production, the former is considered synonymous to '*Reservoir seismic*' and the latter to '*Production seismic*'.

3D High Resolution Seismic

Data Acquisition

3D seismic is recorded over an area in which data is sampled densely along a regular grid and the processed output, available in a volume. On land, 3D acquisition is done with closely spaced grid of shot and receiver points spread over an area called a 'swath'. In the swath layout, receivers are placed on parallel lines and shot points positioned on parallel lines orthogonal to receiver lines (Fig. 1). Shot-point lines are called 'Cross lines' and the receiver, the 'Inlines'. However, there can be several alternate lay out options that can be designed depending on the geological objective.

The survey is called 3D, as its lay-out geometry essentially allows receivers to record reflected waves coming from several azimuthal directions in contrast to 2D data that primarily record reflections arriving from the vertical plane, limited to a single azimuth defined by the source-receiver profile. The close spacing of traces in 3D is characterized by the 'bin' size, which is the minimum area containing the cluster of common mid points (CMP), commonly termed CDP, for stacking. The bin size controls the spatial sampling resolution and typically varies between grid sizes of 12.5×12.5 m to 25×25 m, depending on the dimension of geologic objective to be imaged. Though square-size bins are desirable, often due to operational constrains, rectangle–shaped bins are also used. In marine 3D surveys, data however, are mostly recorded in a set of closely spaced profiles with multi-streamer and multisource (air guns) arrays, towed by the recording seismic vessel for operational efficiency.

The 3D data because of its closely sampled regular-grid permit volume-based superior processing techniques such as surface-consistent static, deconvolution, velocity analysis and in particular the migration process which yield high quality seismic images with improved temporal and spatial resolution. Recording data from multiple azimuths (*MAZ*), help collapse diffractions and out-of-plane events properly and create more accurate three dimensional seismic image of subsurface compared to 2D (Fig. 2). Multiple azimuth (*MAZ*) data also permit mapping azimuth-dependent anisotropy and fractures in formations in the subsurface. Specifically, the migration processes performed in time and depth domain of 3D prestack data play the most important role in enhancing the sharpness and resolution of the images. Prestack depth

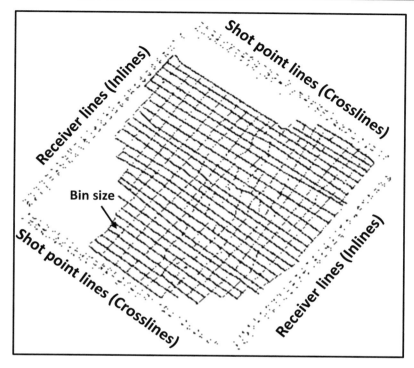

Fig. 1 A typical 3D survey swath lay out showing parallel receiver and source lines, orthogonal to each other. Receiver lines are called Inlines and the shot point lines the Cross lines. The rectangular shaped 'Bin' size is marked by the arrow. Note the regular and close grid geometry for recording data. Breaks in the grid are due to ground logistic problems

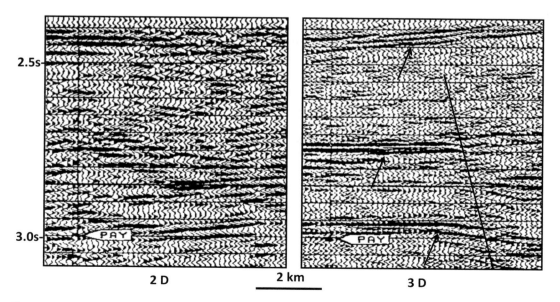

Fig. 2 Example showing 2D and 3D images of the same seismic segment. Note the improvement in clarity and continuity of events and the fault in 3D data (shown by arrow) due to regular and close-grid data acquisition and volume based superior processing. 3D migration plays a major role in reducing noise and improving resolution *(Image courtesy of ONGC, India)*

migration processed in depth domain by using more accurate velocity fields are increasingly in demand and are deliberated below.

Prestack Time and Depth Migration (PSTM/PSDM)

Prestack migration of 3D seismic data is a volume-based operation in contrast to 2D migration which is carried line-by-line in time and depth domain. 3D migrated data in addition to defining better the structural configuration and continuity of geologic features, preserves true relative reflection amplitudes which provides improved analysis of high-end seismic attributes that results in better reservoir characterization.

Prestack Migration (PSTM)

Migration of prestack data in time domain (PSTM) is presently carried out in routine basis because of reduced time and cost, made possible due to faster machines and powerful software algorithms. PSTM data are considered appropriate for interpretation in areas where the depositional and tectonics style are gentle as in simple geologic settings and have little lateral velocity variation. In such situations the PSTM technique assumes constant velocity model valid for vertically varying velocity and usually suffices for structural and stratigraphic interpretation without much ambiguity. But in geologically complex areas where severe lateral velocity variation exists, the time-migrated data (PSTM), which does not consider lateral velocity variations, may be unsuitable. Interpretation of geometry and depth of the geologic features can be erroneous, skewed with uncertainties depending on varying lateral velocity. One would then need depth domain migration (PSDM), which considers lateral velocity variations, even though it is relatively more costly and time taking.

Prestack Depth Migration (PSDM)

Prestack depth domain migration uses laterally varying velocities that take into consideration the transmission (refraction) effects. This is in contrast to PSTM which assumes constant velocity model without factoring for the wave transmission effect, the bending of rays at the interfaces due to refraction. Features in complex geologic settings such as compressional folds and thrust faults, submarine canyon cuts and fills, rapidly and largely undulating bathymetry, salt diapirs and overhangs and shallow carbonate reefs etc., commonly show significant lateral changes in morphology and geometry that cause appreciable lateral velocity variations. In such cases the PSTM fails to image accurately the diffracted and scattered energies by restoring their proper depth positions and necessitates depth migration (PSDM). Given a varying velocity field as input, PSDM effectively removes 'velocity anomalies', for instance, time 'pull-ups' and 'pull-downs' (sags), shifts the geologic events to their correct vertico-lateral subsurface positions in depth, improves imaging and defines more accurately the depth and geometry of structural and stratigraphic features. PSDM in the process also restores the relative true amplitudes of reflections better, which help extract seismic geometric attributes such as coherence, dip and azimuth and curvature more precisely (Chapter "Analysing Seismic Attributes"). Artefacts linked to localized shallow carbonate build-ups and incised valleys are eliminated helping in improved delineation and characterization of reservoirs with better estimates of rock-fluid properties. Precise mapping of faults and their exact edge-terminations and removing fault-shadow related spurious discontinuities lead to improved PSDM coherence slices that can help reservoir and production engineers in monitoring fluid-flow. More importantly, the interpretation of PSDM data makes it more convenient to depth-calibrate seismic with wells and import the seismic-derived rock properties directly into reservoir modelling and simulation work flow platform by seamless integration of seismic with geologic, petrophysical, and engineering data, which are inherently in depth domain.

PSDM Limitations

Though PSDM data has great advantages and preferred by many interpreters, overreliance on it may not be the panacea for all the problems as it also suffers from some limitations. The key to the PSDM migration process is the correct input velocity field, and there lies a paradox, the 'well-known' unknown parameter, 'the velocity'! Irrespective of the method chosen from a host of migration algorithms available in the industry, the PSDM result is only as good as the accuracy of the laterally varying velocity-depth model, the data quality and the structural complexity. Building an accurate velocity-depth model input for correct PSDM output depends on the skill and patience of the individual. It also is governed by factors such as how well the model is constrained with seismic, geologic, petrophysical and engineering data inclusive of anisotropy factor that may require a large number of iteration process. The upshot is, if the velocity model is flawed, depth migration may be at times inferior to time migrated data. More crucially, depths indicated by PSDM data may not render the true depths that tie with the wells.

Case Example

It may be stressed that depth migration and depth conversion are two different processes, the former is an imaging issue while the latter is related to calibration (Etris et al. 2002). The migration stack velocities used in PSDM, despite their accuracy, are not the same as true vertical propagation velocities required for depth conversion because of the horizontal component involved in the former. Consequently, the PSDM depths can lead to errors, at times as high as hundreds of meters between the PSDM and the well depths. An example of a PSDM and PSTM image of a superdeep offshore high amplitude shallow Pliocene seismic anomaly is shown in Fig. 3. It may be noted that comparison of PSTM with PSDM image shows little difference in structural or stratigraphic details. This is because the geologic setting; the target is a shallow simple feature with gradual change in the monoclinal bathymetry with little lateral variations in the overburden velocity, as is exhibited by the regular parallel reflections above the anomaly. Rather on the negative side the PSDM showed depth differences of the order of 15–25 m with the well log depths even at shallow depth and

having low overburden velocity range in the offshore deep-waters (Nanda 2018). Without a priori knowledge, the velocity field input to the depth migration would have been the cause for the discrepancy.

This emphasizes the point that without adequate velocity data to start with PSDM may be ineffective and infructuous even in simple geologic settings. The uncertainties in PSDM depth, however, increases with depth which necessitates many iterations in the tomographic technique for zeroing-on the exact velocity model which even then may not yield true depths because of the reason stated earlier. However, with extra cautious repeated iterations, the discrepancy can be reduced to minimum but may have to be compromised considering its cost-benefit factor.

3D Seismic Interpretation and Evaluation Techniques

Interpretation of 3D volume data is more convenient and permits to comprehend better the stratigraphic and structural styles that are less evident on conventional profile display of 2D data requiring line to line interpretation. Interpretation of 3D data results in more accurate prediction of reservoir geometry and rock-fluid properties for reservoir characterization, needed for field development. However, better-quality interpretation requires unambiguous seismic horizon correlations to start with which requires a meticulous well calibration followed by detailed seismic mapping and evaluation of rock properties.

The 3D interactive interpretation is accomplished fast and accurate by use of powerful and sophisticated interactive softwares. 3D techniques provide several important advantages which are not achievable with 2D data. This includes creation of arbitrary and reconstructed 2D seismic sections in any desired azimuth, display of time/depth slices in plan-view, fault-plane mapping and most importantly, extraction of multitude of seismic attributes that help solve a slew of reservoir and production problems (Chapter "Analysing Seismic Attributes"). Subtle

Fig. 3 Seismic example showing (**a**) post stack time (PSTM) and (**b**) post stack depth (PSDM) migrated images of a superdeep offshore high amplitude shallow anomaly. Note the marginal improvement in PSDM over PSTM. This is explicable as there is no lateral velocity variation indicated by smoothly dipping bathymetry and the parallel-layered overburden strata. The PSDM depths showed discrepancies of 15–25 m with log depths due to flawed velocity input. (After Nanda 2018)

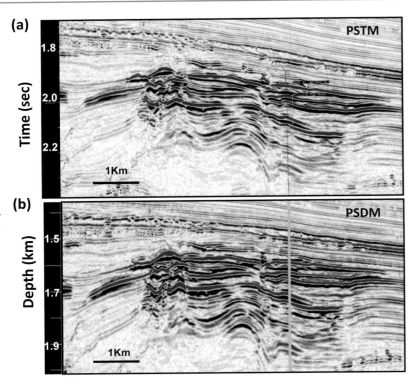

sedimentary features, such as gentle delta progradations or channel cut and fills, often can be seen only on dip or strike lines and may not be perceptible in 2D seismic if the lines are not shot in those specific azimuths. 3D seismic is free of these constraints as an arbitrary line can be generated from the volume data in any direction the interpreter desires to perceive the geological features. Reconstructed or arbitrary seismic lines connecting drilled wells are conveniently simple and can be extremely useful in analysing the relative seismic responses of the reservoir properties in the oil/gas and dry wells. Seismic traverses created through hydrocarbon and dry wells can be excellent displays for reviewing and analysing the variance in seismic response causal to the geology of the wells and set bench marks for seismic prediction of potential hydrocarbon prospects. One of the major advantage is viewing the 3D data horizontally in plan views which reveals subtle stratigraphic features better than when viewed in vertical sections.

Horizontal-View Seismic

One of the most straightforward and particularly effective means of 3D interpretation is horizontal viewing of seismic as against the 2D vertical viewing of sections. Depositional bodies mostly have horizontal dimensions greater than their vertical dimensions and horizontal-view seismic interpretation is likely to resolve small-scale depositional features better in planview (Zeng 2006). Though horizontal viewing of slices cut across the 3D volume of data is widely used for all types of attributes only the amplitude slices are included here. Other attribute slices are discussed in detail in Chapter "Analysing Seismic Attributes" titled "Analysing seismic attributes".

Horizontal-View Amplitude Slices

Horizontal slices cut across the 3D volume displays amplitude variations in planview and provides a convenient quick-look interpretation of seismic data over the entire area. It provides to assess the trend, strike and the extent of the geologic feature in the entire area. Types of seismic slices for horizontal viewing are dealt below.

Time Slice (Horizontal)

Time slice is a horizontal slice cut through the volume at a selected time and is known as 'horizontal slice' or simply 'time slice'. A time slice cuts across different horizons and such slices can be created at regular time interval, usually of 4–8 ms, and can be used for structural mapping of interesting anomalies. Successive time slices increasing in time showing wider extents of anomaly with (time) depth signify anticlinal structures and conversely, synclines if the extent gets narrower (Fig. 4). This is a convenient and quick-look way to broadly map structural or stratigraphic features from number of time slices generated. Time slices are also called 'seiscrop', a term analogous to geologic term 'outcrop' where rocks are studied on a surface traverse. Amplitude time slices are extremely useful as they clearly display discontinuities in planview and are simple and fast ways to map faults (Fig. 5).

Horizon Slice

Horizon slice is a slice cut through a correlated reflection horizon after it is made horizontal known as flattened. Horizon time slice show the behaviour of the horizon, the geologic strata deposited presumably flat. Variations in reflection amplitude provides stratal information along the horizon and facilitates fast and precise mapping of two-dimensional anomalies in the area. The horizon amplitude pattern mapped in the plan-view are easily perceptible for geologic interpretation. While the horizon slices show the stratal properties over the area, the time slices highlight the discontinuities, such as faults. Time or horizon slices are chosen for interpretation depending on geologic objectivity. Horizon amplitude patterns in horizon slices, however, can vary depending on the pick of the reflection phase and precision of horizon correlation and can have different geologic implications (Fig. 6). Because amplitude is sensitive to noise, picking the zero-crossing, the least affected by noise, for horizon slice display may be more reliable.

The slicing technique reveals subtle two dimensional depositional features like channels, deltas, barrier bars, fan complexes etc., better in planview somewhat similar to surface geomorphologic features observed in satellite images. Conventional interpretation of vertical sections may detect these features but the smaller and interesting exploratory objects like point bars, levees, crevasse splays etc., may not be resolved due to limited vertical resolution. An example of a meandering channel, a common exploration play, mapped clearly by horizon slice but quite incomprehensible on a vertical section, is shown in Fig. 7.

Time slices (Seiscrops)

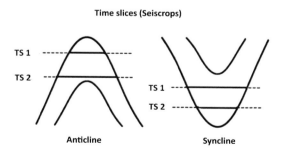

Anticline   Syncline

Fig. 4 Schematic illustrating concept of 'time slice' on planview for an anticline and syncline. TS1 and TS2 denote horizontal slices (time slices) cut across the 3D volume data at two time levels. For anticline the slices would be viewed as of wider extent with increase in time whereas for syncline it is the reverse, narrower with time. Marking the extent of the segments at each time slice makes the structure map for the feature

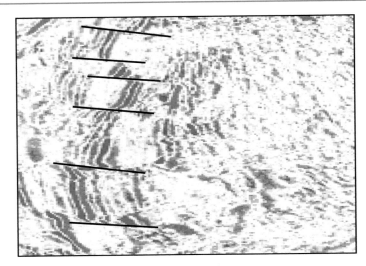

Fig. 5 Example of a 3D seismic amplitude time slice depicting clearly the faults (marked) and their geometry, extent and orientation over the entire area. The fault displacements can play a major role in impeding connectivity within a reservoir and impact production (Image courtesy: Satinder Chopra, Calgary)

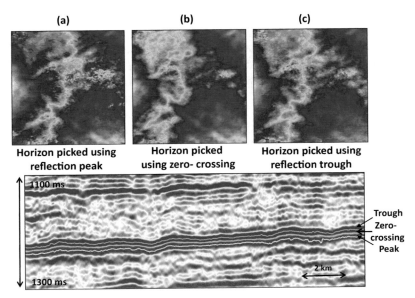

Fig. 6 Illustrates the impact of choice of reflection phase on display of horizon slices in planview. Horizon slices are horizontal slices of a horizon after it is flattened. The lower panel shows vertical seismic segment and the above panel the horizon slices cut through (**a**) reflection peak, (**b**) reflection zero-crossing and **c** reflection trough of the horizon marked. Note the significant variations in the horizon slices due to noise which can have major impact on geological interpretation. Zero-crossing has least noise and may be preferred for reliable horizon slices (Image courtesy: Satinder Chopra, Calgary)

Stratal Slice

Stratal slicing is a variation of horizontal slice and a technique which is commonly and widely used. Horizon slice being a slice along a bedding plane (horizon) essentially represents the depositional surface of a feature and works well

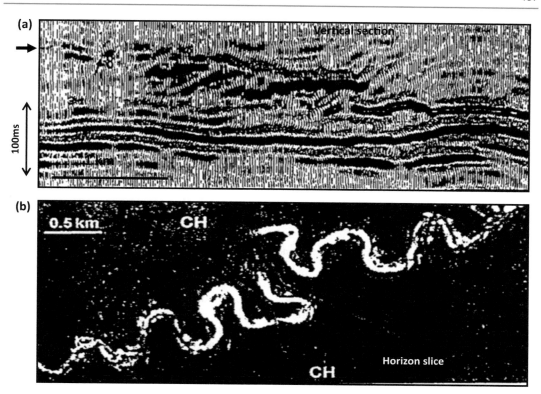

Fig. 7 Showing example of a seismic vertical section (**a**) and a horizon slice at the level indicated by the arrow (**b**). Note the clear image of a meandering channel revealed by horizon slice displayed in planview that can be easily mapped and is incomprehensible in (**a**) the vertical section (Modified after Kolla et al. 2001)

mostly in conformable sequences which are presumed deposited flat, as in 'layer-cake' geology. Horizons are therefore flattened before slicing so that the slice does not include feature belonging to another geological age as it happens in time slices. Scanning through parallel horizon slices at close time intervals in a formation reveals the vertical and lateral changes in a depositional sequence and works well when it is of uniform thickness. Nonetheless, where geologic sequences change thickness laterally, as are often found in nature, the horizon slices may sample diachronous events of different geologic age. In such cases, to limit slicing along the bedding surfaces, geologic time surfaces (stratal surfaces) are constructed from the seismic volume by dividing the variable time interval between two seismic reference events into a number of uniformly spaced subintervals. Slicing along these surfaces is known as 'stratal slicing'

or 'proportional slicing' (Zeng 2006) and is likely to provide more details on variability of facies within the sequence. Seismic attributes mapped from the stratal slices can then be analysed in terms of depositional systems.

Window-Based Amplitude

Window based amplitude is another technique usually adopted to display amplitude values computed in a specified window of seismic stack data. Essentially, the amplitudes for all samples in a selected window are considered for computing average, maximum, or RMS (root mean square) amplitudes which can be chosen for display in planview. The average amplitude computes the mean of amplitudes, whereas, the maximum computes the maximum of the absolute value of peak and trough of the amplitudes in the window. The most commonly practiced RMS amplitude is the computed value as the square root of the sum of squared amplitude values divided by the number of samples within the specified window. Squaring offers the amplitudes to stand out best though, it is also more prone to show the unwanted weak events and noise. The windowed amplitudes are basically used as a simple and quick means to identify the leads as prominent potential DHI amplitude anomalies and estimate hydrocarbon resource potential of the geologic plays in the window over the volume of data. (Fig. 8). The horizon correlation and the selection of the window, which is usually of the order of 20–50 ms, is important as varying windows would display different amplitude patterns implying diverse geological implications. Window must be chosen carefully, encompassing only the interesting features along with the type of amplitude to be computed. Selection of a broader window with computed RMS amplitude is likely to overestimate the resources.

In clastic depositional set-ups amplitudes are often helpful in delineating hydrocarbon sands for which appropriate type of slice must be chosen for the purpose. The relative advantages and limitations of each method may be weighed based on the specific geologic issue at hand. For instance, window-based RMS amplitude may

work for simple stratigraphic reservoirs but not for multiple thin reservoirs occurring at different stratigraphic levels within a broad window that is chosen arbitrarily. For instance, consider the offshore channel-cut and fill complex, a typical exploration play, for quick evaluation shown at Fig. 8. A chain of discontinuous thin channel sands imaged as high amplitude anomalies can be seen in a broad window of about 100 ms and if the window is chosen for calculating RMS amplitudes to represent hydrocarbon it would lead to a highly exaggerated inflated resource estimate. Care must be taken to restrict correlation of the events in a small window of 20–25 ms, either correlated by their external form or some such other criteria to restrict the window to evaluate the plays. Horizon and stratal slices, on the other hand are less impacted by unrelated geologic events and are preferred for delineating reservoirs provided the reflection horizon phase is reasonably identified and tracked.

The efficacy of the horizontal slices are best realized when shown riveted to their corresponding position in the vertical sections. This is called 'chair display' and by cross-referring both the vertical and horizontal slices, it brings out clarity and confidence in interpretation of geologic bodies (Fig. 9). Resolution of stratigraphic

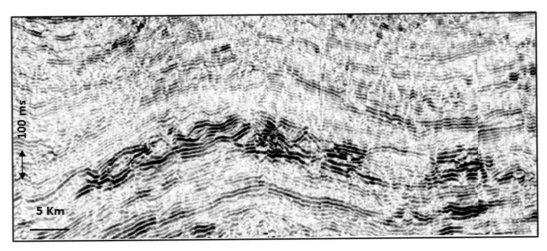

Fig. 8 Seismic image of normal full-angle stack offshore seismic segment viewed in vertical section showing a chain of high amplitude anomalies of 'channel cut and fill' features in a window of about 100 ms. The channel

cut and fill sands are discontinuous belonging to different deposits and choosing the broad window for computing RMS amplitude to estimate resources may be flawed. (Image courtesy, ONGC, India)

vertical time section   **vertical time section**

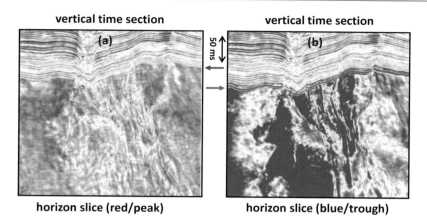

horizon slice (red/peak)   **horizon slice (blue/trough)**

Fig. 9 Illustrates chair-display of attributes, horizon amplitude slices in plan view (*bottom* part) correlated to the vertical section on *top*. Horizon slices shown at two levels (**a**) red horizon and (**b**) blue horizon shown by arrows in the vertical section. Though the reflection horizons look parallel in vertical sections, the plan view amplitude maps over the area show noticeable differences which may have geologic significance. (Images courtesy of Arcis Seismic Solutions, TGS, Calgary)

prospects are best achieved by using both vertical and planview sections in conjunctions and 'chair displays' are extremely useful for the purpose.

Horizontal-View Seismic for Sequence Stratigraphy Interpretation (SSSI)

Traditionally, conventional interpretations are carried out on vertical seismic sections as the image is expected to replicate the subsurface geology in depth. But with emerging advantages of seismic 'horizontal-view' techniques, it has evolved as a powerful and fast technique for seismic sequence stratigraphy interpretation (SSSI) of 3-D volume data and the work flow is briefly outlined.

SSSI Framework: Horizon Cube

The seismic sequence stratigraphy interpretation by horizon-viewing of seismic is essentially based on two-step workflow, creation of a 'Horizon Cube' and its transformation to 'Wheeler domain'. 'Horizon cube' is a dense set of correlated stratal surfaces, each interpreted to represent a relative geological age. Major sequence boundaries are mapped and all possible reflection events within it are auto tracked to create a large number of horizons. Auto tracking can be either model based or data driven. In the former mode, tracking is done interactively with a geologic model by fixing or interpolating horizons parallel to upper/lower boundaries (Brouwer et al. 2008). In the data-driven mode it deploys auto-tracking by following dip-azimuth of the events. Essentially, each horizon created corresponds to a stratal surface and assigned a geologic time, the stratal surfaces representing the chronostratigraphic events.

SSSI Framework: Wheeler Domain

The stratal surfaces created in the Horizon Cube are flattened and the data transformed into what is known as the 'Wheeler domain'. Time slices in the 'Wheeler domain' are the equivalent of the horizon slices in conventional seismic. The Wheeler transformation is an extremely convenient way of graphic display technique for better comprehension of chrono-stratigraphic studies (Fig. 10). The Wheeler time slices make it easier to visualize and interpret spatial distribution and timing of sediment deposition.

Volume based SSSI interpretation techniques, because of fast and dependable evaluation of geologic basins, has become part of regular workflow in many companies, especially dealing with large volumes of 3D data. However, the techniques work well in simple geologic set-ups where sediments are not much distorted by tectonics. It also requires good quality data without

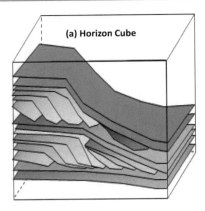 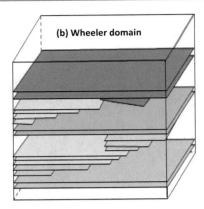

Fig. 10 Schematic illustrating the two-step work-flow of seismic sequence stratigraphy interpretation (SSSI) of 3D volume data. (**a**) The seismic 'Horizon Cube' showing the correlated horizons and (**b**) transformation of events to 'Wheeler domain'. Wheeler domain time slices are the horizon slices of correlated seismic reflection horizons. Wheeler domain color display facilitates better interpretation of spatial distribution, sequencing of beds and their depositional environment (After Brouwer et al. 2008)

noise, suitable for auto tracking which though accurate, may not be correct in many instances.

3D Volume Interpretation and Visualization

The higher resolution and densely sampled 3D volume data proffer an excellent opportunity for volume based interpretation utilising 3D visualization softwares without getting into the customary interpretation done profile wise. Interpretation of data and visualization is linked to mind game which helps perceive and realize better the images of subsurface geologic features. 3D volume visualization technique achieves this by enhancing optimal depiction of images and interpretation outcomes by integrating these with powerful computer graphics, image processing, computer vision, colour blending, signal processing and eventually displaying on large and high resolution screens for the mind to grasp and perceive subtle features.. Visualization tools make interpretation faster, convenient, inclusive and reliable that leads to mitigation of uncertainties in the risky ventures of hydrocarbon exploration and production.

Notwithstanding the huge advantages of 3D volume interpretation technique, many interpreters still continue to practice 2D section-based traditional interpretation of 3D volume data. Volume interpretation and visualization process requires creation of an efficient multidisciplinary geodata base for seamless manoeuvring and integration of geological, geophysical, geochemical, petrophysical and engineering data. Visualising and evaluating comprehensively the maze of enormous multidisciplinary data can indeed be challenging. The system would also require sophisticated hardwares, softwares and attendant infrastructure including capable and skilled analyst to be deployed and maintained that could be highly expensive.

3D Seismic Applications

3D seismic is primarily used in the petroleum industry for reservoir delineation and characterization, described in details later in the chapter. However, other important applications, worth mentioning, are prediction of pore pressures and shallow hazards in offshore drilling for conducting safe and economic operation.

Pore Pressure Prediction

Problems encountered in drilling are often the overpressures, blowouts, borehole instability, mud loss, borehole collapse and stuck pipes, which besides being hazardous, cause more operational time and cost especially in offshore. Prior knowledge of pore pressures not only helps alleviate the hazards but also optimize drilling by planning proper mud weight and casing policy, to achieve well completions in time.

Pore pressures are generally calculated from the well data (Chapter "Seismic Wave and Rock-Fluid Properties") at the borehole but difficult to be predicted in places away from it. High pore-pressure (overpressured) formations show velocities and impedances lower than normally pressured formations at the same depth. 3D data, using these properties can predict formation pore pressures over the area. Simply stated, pore pressure can be determined if the overburden and effective pressures are known. The overburden pressure is obtained from well data. Effective pressure affects elasticity and consequently the seismic properties and can be estimated from the seismic interval velocities. There are several empirical relations to compute pore pressure but the most commonly used is the Eaton's method relating interval velocity to pore pressure (Chopra and Huffman 2006). The prediction reliability therefore depends on the accuracy of the seismic interval velocities used. Seismic interval velocities are highly sensitive to the way they are estimated during data processing and can be done in several ways, discussed in detail by Chopra and Huffman (2006). Improved 3D data quality and processing algorithms enable determining reliable velocities such as from horizon velocity analysis, reflection tomography based depth migration and seismic inversion that are used to estimate effective pressure leading to prediction of pore pressures.

However, the above method requires well data for pore pressure estimation. With no well data information, such as in virgin areas, high pore pressure can be indicated from seismic data as caution for the drilling engineers. Overpressured formations, because of low effective pressure and near-constant velocity (Chapter "Seismic Wave and Rock-Fluid Properties") are generally imaged as poor reflection zones (Fig. 11a). This can be corroborated by NMO velocity analysis; while increasingly higher velocity picking with increasing depth is normal, for overpressured sections the velocity will show constant or reverse trend (Fig. 11b). This can be also supported by seismic facies analysis from seismic stratigraphy which indicates depositional environments and the facies likely to be overpressured.

Offshore Shallow Drilling Hazards Prediction

Locations of upcoming drilling wells are checked for presence of shallow hazardous zones that could imperil offshore operations. Presence of soft, loose and mobile strata at or near the bottom and shallow high pressured gas charged pockets can endanger safe installation and drilling operations. It is mandatory by maritime law that the sea floor and the sub strata around the target area be assessed for these hazards for safe drilling and other related operational activities. Several surveys such as sea bottom sampling, shallow soil coring and high resolution acoustic profiling by 'sparker' surveys are available to locate the hazardous zones.

Marine 3D shallow seismic data can be acquired to image the sea bottom and the sub-bottom zones, which exhibit adequate resolution that allows detecting and mapping features linked to potential drilling hazards. Buried river channels, clay 'dumps', localised gas pockets and seepages in the sub bottom constitute a few of these hazards which endanger safety of drilling rigs and ships to operate. Mobile clay dumps and channel-fills are highly unstable and can collapse during erection and operation of jack-up rigs making it perilously risky (Fig. 12). Gas seepages through sea bottom, phenomena often

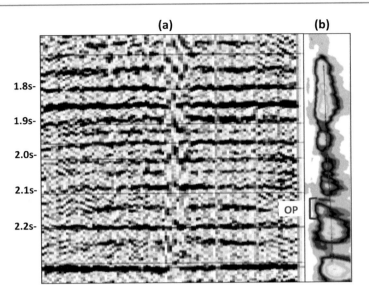

Fig. 11 Example illustrating indication of overpressured zones in NMO velocity analysis for stack. (**a**) Stacked seismic segment and (**b**) the NMO velocity analysis panel from which velocities were picked for stacking. Velocity picks in the lower part (**b**) show trend as constant, without increase with depth, a typical indication of overpressured section. Also note in this part the velocity pick showing a reversal (marked OP), suspected overpressured zone (after Chopra and Huffman 2006)

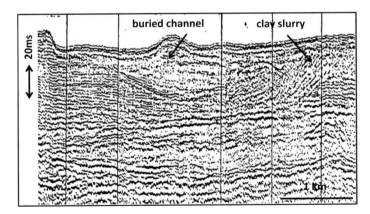

Fig. 12 Example of 3D seismic segment showing potential shallow drilling hazards in offshore. The 'Buried channel' and 'Clay slurry' marked by arrows are extremely pliable and mobile posing risks to installation and stability of jack-up rigs for safe operation. (Image courtesy ONGC, India)

observed especially in offshore deep- water areas, reduce buoyancy of water and can impede deployment of semi-submersible floaters at the site for drilling (Fig. 13). Likewise discrete, high-pressured shallow gas pockets, gas charged water sands and shale can also cause blow-outs or shale flow into well, posing drilling hazards, time delays and other related troubles (Fig. 14).

Reservoir Delineation and Characterisation

By far the most significant interpretive advantage of high density-high resolution 3D volume is the ability for quick and accurate analysis of multitude of seismic attributes extracted from the data (refer Chapter "Analysing Seismic Attributes").

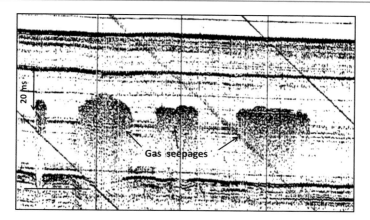

Fig. 13 Showing shallow marine seismic image of shallow offshore gas seepages that can be potential drilling hazards. Gas leaking through sea bottom reduces buoyancy of water, endangering operation of semi-submersible floaters deployed for drilling. (Image courtesy ONGC, India)

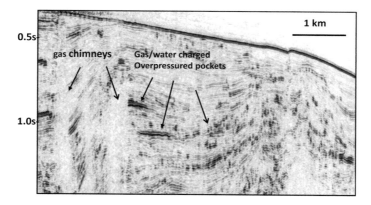

Fig. 14 Example of an offshore seismic segment imaging shallow drilling hazards. Note the strong amplitude events suggesting gas/water-charged high-pressured pockets. The s transparent vertical features cutting through the strata are inferred gas chimneys that are potential drilling hazards. (Image courtesy, ONGC, India)

The attributes help estimate the reservoir properties to evaluate and estimate the in-place reserves in exploration stage and more accurately for delineation and characterization of reservoir during development stage. Vital reservoir parameters estimated from seismic, in conjunction with well data are usually the prime input needed for initial static reservoir modelling and consequently planning production profile. 3D seismic is indispensable in the gamut of exploration and production as crucial decisions are taken mainly based on seismic interpretation and evaluation.

Reservoir delineation and characterization are terms, often used synonymously, may be distinguished as separate issues, though they are co-linked. The essential elements that fully describe a reservoir are:

(i) Reservoir geometry (shape, size and thickness)

(ii) Depth to reservoir top

(iii) Fluid contacts (GOC, OWC, OSC, GWC, etc.)

(iv) Rock-fluid properties (porosity, permeability, fluid saturation, pore pressure etc.)

(v) Reservoir heterogeneity (facies change, barriers, faults and fractures etc.).

The first three aspects deal primarily with the reservoir description and may be subsumed under 'reservoir delineation', while the last two are headed under 'reservoir characterisation', which involves the reservoir rock and fluid properties. After hydrocarbon find, typically, the reservoir needs to be fully delineated and its productivity assessed by drilling additional wells, known as delineation/appraisal wells. Reservoir delineation essentially deals with depth mapping of the prospect with fluid contact to provide volume of hydrocarbon bearing rocks, required for assessing in-place hydrocarbon. Reservoir characterization follows with estimating the reservoir properties more precisely to build initial reservoir models and plan production profiles.

Reservoir Delineation

Hydrocarbon discovery is followed by reservoir delineation to assess production potential of the prospect and is a practice particularly in offshore where exploration is cost-intensive and needs more and reliable information to reduce risk. To optimally locate delineation wells for drilling, a fresh set of structural and reservoir facies maps may have to be prepared. Mapping the reservoir involves mainly three steps, (i) well calibration to pick proper phase and polarity of target horizons, (ii) correlation and lateral extent and finally (iii) seismic time maps converted to depth. Availing maximum benefit of superior 3D

images with much improved vertical and lateral resolution warrant stringent quality check for the above three steps, notwithstanding their limitations (Chapters Seismic Interpretation Methods and Borehole Seismic Techniques), to deliver precise definition of reservoir geometry. The above steps in the interpretation work-flow may appear elementary but their outcomes are extremely significant in the exploration endeavour. Experience shows that if small thing are taken care of in the initial stage, bigger problems will be solved by themselves to a large extent in the later stage.

Seismic-Well Tie and Horizon Correlation

Seismic tie with the well, is the first step which decides picking of the exact reflection phase (peak/trough/zero inflection) corresponding to the reservoir top and bottom that require correlation. Picking correct phase is important and acts as a guide for horizon correlation and its lateral extent by reflection character. Horizon correlation based on reflection rather than a customary time tie in the loop, shows not only the precise extent of lateral continuity to map the reservoir limit but also indicates rock property changes laterally (Fig. 15). Often, the well calibration faces problems especially in land data, leading eventually to flawed picking and correlation of reflection horizon. Phase-picking can be crucial as a particular peak/trough may show variance in lateral continuity due to frequency split-ups, significantly impacting estimate of the

Fig. 15 3D seismic segment showing change in reservoir top reflection character (arrow mark) indicating reservoir facies change and delineates the limit of the reservoir by its lateral extent. *(Image courtesy of ONGC, India)*

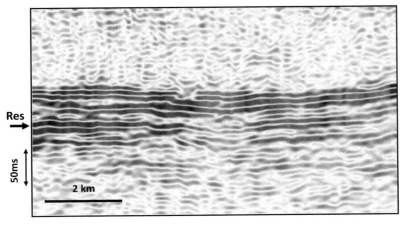

prospect volume, often one of the major risk factor in volumetric estimate.

Depth Maps

Detailed structural and stratigraphic interpretation delivers accurate seismic structure maps providing depths to top and bottom of reservoir, its thickness and lateral extent. Fluid contacts are also factored in estimating hydrocarbon bearing rock volume. Different kinds of map may be prepared as per requirement depending on the type of trap and fluid contact. For instance, in case of structural traps with the fluid contact (OWC/OGC) occurring within the reservoir thickness, known as 'bottom water' (Fig. 16a), the hydrocarbon bearing rock volume can be calculated from the top of reservoir map and the water contact. This necessitates the structure map at the top reservoir to be accurate to portray the amplitude of the structural closure and areal extent. A map of reservoir bottom may not be necessary. Whereas, for traps with fluid contact occurring in the flanks, within the reservoir thickness, known as 'edge water' (Fig. 16b), volume calculation would need accurate maps at both top and bottom of the reservoir. For stratigraphic traps where the contact is edge-water (Fig. 16c), the volume calculation is similar except for that the zero-thickness on the isopach map, representing the updip limit of reservoir be meticulously demarcated. Isopach map denoting

the hydrocarbon column (thickness) is important and because it is sensitive to error due to inaccuracies involved in mapping two horizons, it requires careful attention.

In some cases, the gross hydrocarbon column encountered in the well may be greater than vertical closure of the structure mapped from seismic. If the seismic map is accurate, it may signify multiple reservoirs that need different set of maps to delineate; however, the reservoir bottom maps may not be needed in such cases. For shale contacts (OSC/GSC), where the reservoir is immediately underlain by shale, it is difficult to estimate the exact rock volume unless a down dip well is drilled to establish the water contact. The seismic interpreter may have to look into these aspects to be sure that suitable maps are prepared and they honour the data encountered at the well(s). Needless to add that velocity function is the most important factor and that due diligence must be done to check its accuracy (Chapter "Seismic Interpretation Methods").

Volumetric estimates, taking into account the structural amplitude and the closure area signifies maximum possible hydrocarbons-in-place considering hydrocarbon fill up to the spill-point of the structure (see Chapter "Seismic Stratigraphy and Seismo-Tectonics in Petroleum Exploration"). The whole of the reservoir may not be productive and depending on reservoir porosity only portion or portions of the reservoir is considered capable

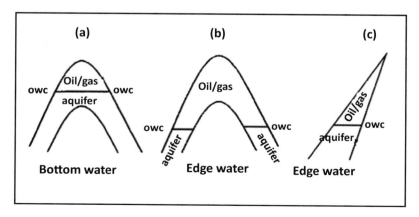

Fig. 16 Schematic illustrating types of hydrocarbon-water contacts in reservoirs. (**a**) Bottom water when the contact is below in the entire reservoir and (**b**) edge water when it is in flanks of structural traps and (**c**) edge water in stratigraphic traps

to produce, which are termed the 'pay'. Ideally it would be best if the pay level structure maps can be prepared from 3D data. But in the absence of mappable reflections from the pay, sometimes the reservoir top and bottom are mapped and annotated as pay top and bottom maps which could be misleading. Some interpreters use the term 'near pay' which may describe the structural aspect, but may not truly represent its aerial extent and more importantly the reservoir characteristics away from the well. This is because the reflection phase mapped is different from that of the 'pay' and may not define the true properties of the pay horizon.

In the context it may also be mentioned that the true stratigraphic thickness (TST) of the pay be considered for reserve estimates instead of true vertical thickness (TVT) used in computation. True stratigraphic thickness of a bed is the thickness measured in the direction orthogonal to bedding plane representing the thickness of the bed at the time of its deposition (paleo thickness), whereas, true vertical thickness is the thickness measured along the plumb line. In thick and highly dipping large reservoirs the calculated volume difference can be sizable. The interpreter needs to be aware of inclined well depths reported that were logged which is the bed thickness measured along the bedding plane (MT). This is usually converted to true vertical depth (TVD) to be compatible for seismic calibration. The true stratigraphic thickness (TST), the true vertical thickness (TVD) and the measured thickness (MT) in a deviated well are shown in (Fig. 17). Seismic-well calibration in such deviated wells may also be tricky; though the drilled depths are corrected to TVD, the log response of rocks along the traverse in the deviated borehole may be different from that at the location vertically below the drill-point, which the seismic trace represents as the subsurface location.

The newly prepared detailed structure maps are also vital for deciding appropriate locations for drilling delineation/appraisal wells. For instance, in fault-bounded, steeply dipping structure, saturated with light oil, it is important that the early production wells are suitably located close to the highest structural position so that it gets the benefit of active gravity drainage for maximum primary recovery. Starting earlier production from wells, located structurally lower on the flanks, may run the risk of leaving behind oil at the highest part of the crest, known as *attic oil*, which is difficult to produce economically later on. This requires accurate structural mapping including the faults. However, velocity estimation for depth conversions at this stage poses a stiff challenge due to lack of adequate well velocity data and prediction errors can result in the well ending up in water if the actual reservoir top encountered happens to be deeper than predicted.

Reservoir Characterisation

Characterisation of a reservoir deals with quantifying its rock and fluid properties, such as the porosity, shaliness, permeability and hydrocarbon saturation. Reservoir heterogeneity is another important factor that also needs addressing. These information are obtained at the delineation/appraisal wells, but need to be known in the interwell space and away from the wells for building the reservoir model for the prospect. 3D seismic constrained with well data provide these parameters and help characterize the reservoir.

Porosity

Porous sedimentary rocks are known to have lesser density and bulk modulus and exhibit lower seismic properties such as velocities and impedances. Lowering of velocity is generally considered an indication of increased reservoir porosity provided the lithology remains the same. Similarly, amplitude changes of reservoir top reflection which depends on impedance contrast can be indicative of lateral change in porosity if the overlying rock presumably remains unchanged. Porosities can be estimated quantitatively from analysis of attributes like amplitude, velocity and impedance studied from seismic data. Generally, the reservoir cap rock which is mostly shale, does not change laterally over the

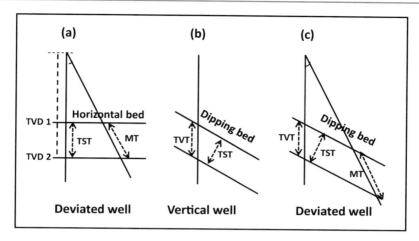

Fig. 17 Schematic illustrating true stratigraphic thickness (TST), true vertical thickness (TVT) of beds (reservoirs) in vertical and deviated wells. MT represents the thickness of bed measured during drilling and logging in deviated wells. Note the TVT, commonly referred as TVD, may need correction for exact thickness of dipping reservoir (pay column) which increases with dip and can impact reserve estimates particularly for large steeply dipping reservoirs

prospect and the lateral changes in seismic amplitude as perceived in extracted stratal slices can be attributed to reservoir porosity. However, the reservoir facies may change laterally, which often is the case and amplitude variations can then only qualitatively denote relative porosity changes based on consideration of other geologic factors. For instance, clastic reservoir's proximity to source may indicate coarser sands and better porosity.

However, as stressed earlier, the seismic calibration needs to be precise to provide proper benchmarking set by the measured porosity values at the wells, so as to ensure reliability of seismic prediction in prospect scale. Sometimes the measured porosity values at the wells, duly constrained by seismic and the geology of the area, are geostatistically mapped to indicate porosity variations over the prospect using techniques like kriging or cokriging. With good quality data, spatial distribution of porosity values at inter-well regions and beyond can be reliably estimated from inversion of seismic data. Seismic inversion, a technique which transforms seismic reflectivity to reservoir rock and fluid properties is discussed in Chapter "Seismic Modelling and Inversion".

Shaliness

Most reservoirs, in particular clastic reservoirs contain some volume of shale with clay minerals, referred as 'V_{clay}' by petrophysicists. Shaliness influences seismic properties and needs to be considered for reliable prediction of reservoir properties. Shales occur in several modes in the matrix and their effect on bulk density and porosity in a reservoir is shown (Fig. 18). Shales occurring in matrix as thin layers lower both the matrix density and porosity and may perceptibly affect seismic response (Fig. 18c). More about this is discussed in Chapter "Shear Wave Seismic, AVO and Vp/Vs Analysis".

Permeability

Permeability is the property of a porous rock which describes the ease with which a fluid passes through the interconnected pore spaces. Permeability depends on effective porosity (interconnected pores) but is not the same as permeability also depends on the pore network geometry, such as pore throat and tortuosity. Permeability is calculated from well log data or by laboratory measurement of core. This is the most important parameter in reservoir characterization, but unfortunately permeability does not

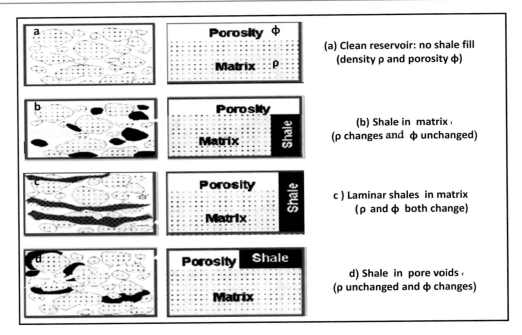

Fig. 18 Schematic diagram showing types of shale fills in reservoir matrix which affect density and porosity (annotated). Note the occurrence of laminar shales (**c**) in the reservoir which lowers both the matrix density and porosity may cause perceptible seismic response. Shale contents are corrected in petrophysical computations for input in reservoir characterization

directly influence seismic properties. Permeability can be linked to effective porosity but strictly not to porosity estimated by seismic though in most cases it is found to be related. Reservoir thickness also impacts permeability and the estimated pay thickness and porosity from seismic data, combined together, may indicate a more reliable measure of permeability.

Fluid Saturation

Fluid saturation affects seismic properties variously depending on type of fluid and its volume fraction in the pore space. (Chapter "Seismic Wave and Rock-Fluid Properties"). In general, sedimentary rocks saturated fully with liquid tend to show increase in compressional velocities and slight decrease in shear velocities due to increase in fluid density. Most hydrocarbon reservoirs are partially oil saturated and the effect on seismic may not be perceptible. Gas, on the other hand shows considerable change in seismic properties which is not dependent on saturation (Chapter "Seismic Wave and Rock-Fluid Properties").

Permeability and saturation, the two more important parameters are usually difficult to determine from 3D seismic, but may be qualitatively assessed by an experienced interpreter from a synergetic analysis of seismic, geological and well data. (Chapters "Shear Wave seismic, AVO and Vp/Vs Analysis" and "Seismic Modelling and Inversion".)

Reservoir Heterogeneity

Most reservoirs are heterogeneous for reservoir engineering purposes. Heterogeneity in a reservoir can be attributed to facies changes, faults and fractures causing anisotropy and other anomalous complications in hydrocarbon flow during production. Evidences of reservoir facies variations in inter-well areas can be picked from seismic from dissimilarities in their reflection character which may be attributed to reservoir heterogeneity. Seismic wiggle and variable area mode of display is better suited for viewing reflection character based on amplitude and waveform shapes to detect heterogeneity (Fig. 19). Presence of high permeable layers and fractures, vertical and lateral barriers in the form of faults or shale streaks can cause '*flow units*' detached within the reservoir. '*Flow units*' are the portions of the reservoir that have properties of consistent fluid flow. A reservoir may have several flow units depending on degree of heterogeneity and they may not be in communication with each other. Reservoir continuity and reservoir connectivity are two different but important aspects of fluid flow; reservoir continuity does not necessarily assure connectivity which may be impeded by heterogeneity. Without this being factored in initial reservoir simulation model, it can lead to unforeseen anomalous fluid-flow patterns during production. Volume-based seismic facies and multi-attribute analysis (Chapter "Analysing Seismic Attributes") with accurate structural maps are useful to identify the various elements such as fractures and anisotropy accountable for causing heterogeneity, which can then be factored in for more accurate reservoir modelling.

The areal extent of the reservoir, the hydrocarbon thickness (pay), porosity and saturation constitute the key seismic inputs to initial reservoir modelling, known as static modelling.

Reservoir simulation for fluid flow patterns, however, need other properties like viscosity, permeability, bubble point, capillary pressure, and pore pressure as inputs, and are derived from log and engineering data measured at the wells.

4D (Production Seismic)

4D seismic is essentially a time-lapse 3D survey, repeated after a period of time following production from a field and is considered a useful tool for reservoir monitoring. During production stage, the virgin fluid saturation and pore pressure decrease with depletion. This increases effective pressure causes change in rock-fluid properties which impacts fluid flow pattern. These changes in rock properties can be detected by a time-lapse repeat 3D survey. Simply stated, the differences in seismic attributes of two 3D surveys, the initial survey before production, normally called the 'baseline' and the repeat survey made after a period of production, referred as 'monitor', is skilfully exploited to address the dynamic changes in the reservoir properties due to production.

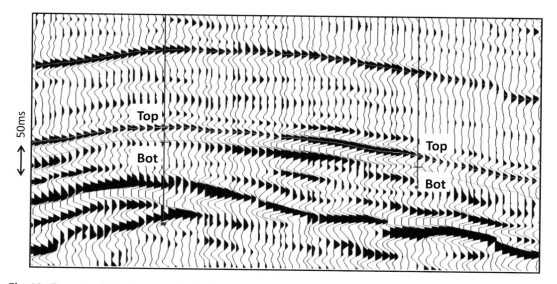

Fig. 19 Example of seismic segment indicating reservoir heterogeneity. Note the big dissimilarity in amplitude and waveform of reflection from *top* and *bottom* of the reservoir between the two wells indicating facies change and linked heterogeneity in the reservoir. Reservoir *top* and *bottom* identified from log is marked on seismic. (Image courtesy of ONGC, India)

Because the differences in the two sets of observed seismic properties are linked directly to the changed rock-fluid parameters, it is necessary that the data acquisition and processing parameters of the repeat seismic campaign be ideally the same as that of the baseline survey. This is usually difficult to achieve in real practice because of several factors, environmental and operational which include the change in ground conditions, geophones and recording instruments and the ambient background noise. These shortcomings are to somewhat resolved by using special sophisticated softwares that attempt to normalise amplitude, frequency, and phase and the bin locations in the two data sets so as to bring them on to a common platform for comparison. Interpretation and evaluation of 4D data is mostly based on attributes, similar to that in 3D, such as amplitudes, P- and S-impedances and Vp/Vs ratios derived from seismic elastic inversion. Powerful and efficient image display graphic softwares are generally used for detecting the subtle differences in seismic properties for analysing the changes in reservoir parameters and fluid flow behaviour occurred during production. 4D is used as a seismic reservoir monitor (SRM) which besides monitoring fluid production behaviour also provides the altered reservoirs parameters for dynamic reservoir characterization and is considered synonymous to 'Production Seismic'.

Seismic Reservoir Monitoring (SRM)

4D seismic is generally carried out after a few years of production and when anomalous fluid flow patterns are noticed. Seismic reservoir monitoring helps make necessary course corrections in the production plan by characterizing the reservoir with altered rock properties and is an important aspect of reservoir management, which is optimal production with minimum cost. Though several seismic attributes can help monitor reservoir, the discussion here is limited to mainly the use of amplitude. 4D seismic provides valuable help on several aspects of reservoir management which include:

(i)   *Dynamic* reservoir characterization
(ii) Reservoir drive types, water/gas coning and gas phase formation - Indicator
(iii) Detect areas of bypassed hydrocarbon (oil)
(iv) Detect permeability barriers and high permeable pathways
(v)   Monitor secondary and enhanced oil recovery (EOR) efficiency
(vi) Reservoir geomechanics, anomalous fluid production.

Dynamic Reservoir Characterisation

Initial parameters provided for reservoir modelling and flow simulation is referred as the static reservoir characterization. A field under production requires reservoir monitoring which is basically about observing the fluid flow patterns over a period of time to evaluate its consistency with the predicted production profile planned to achieve optimal reservoir management. This is known as history matching and may have to be repeated several times during production. Often, in major fields, the actual production pattern may not follow the initial planned profile because of the changes in rock properties and reservoir parameters. The inconsistencies noticed in production behaviour are analysed and remedied by redoing the static reservoir modelling with changed rock-fluid parameters. This is known as dynamic reservoir characterization. Fresh fluid-flow simulation with dynamic reservoir characterization may need production plan to be tweaked with mid-course corrections and suitable modification of future drilling plans. Field monitoring is a continuing process of matching actual with predicted production and models may have to be redone from time to time, particularly in large offshore fields and injection and production wells accordingly planned.

Reservoir monitoring process can be assisted by time-lapse 3D (4D) seismic. The reservoir properties are usually modified by new and additional information from the newly drilled wells or by reinterpreting existing well and the production data with passage of time. But there can be uncertainties about the reservoir rock and fluid parameters, especially in the regions between and away from the wells. Calibrated 4D seismic attributes, constrained with log, core analysis and engineering data can assist filling in the inter-well areas with the new reservoir parameters estimated from 4D. Nonetheless, the inputs from seismic need to be authenticated by reservoir simulation and must be consistent with all other information from multiple data sets like geological, well logs and engineering data.

The important reservoir parameters for dynamic reservoir characterisation are the porosity, saturation and relative permeability. Seismic impedance is related to lithology, porosity and fluid saturation in a reservoir. Primary recovery of hydrocarbon over a period, results in drop in fluid saturation and the reservoir exhibits lower pore pressure. This increases the effective pressure which may lead to compaction of reservoir and lowering its porosity, especially in unconsolidated sands and can create perceptible changes in seismic impedance due to changes in geomechanical properties of reservoir rock (Chapter "Seismic Wave and Rock-Fluid Properties"). The differences in 3D and 4D seismic impedances may thus be related to porosity and saturation change, presuming no change occurs in other properties. Similarly, fluid saturation is linked to pore pressure and the change due to fluid production can be estimated from 4D seismic, though the impacts of individual parameters, porosity and fluid saturation, is difficult to be discriminated for quantification. As, earlier mentioned estimating gas saturation from seismic can be a problematic issue (Chapter "Seismic Wave and Rock-Fluid Properties"). However, the most crucial parameter controlling flow simulation and production is the relative permeability, which is the ratio of effective permeability of a fluid at a particular saturation to its absolute permeability at total saturation. Obviously, the relative permeability varies as saturation decreases with production and is a key factor that needs attention in dynamic characterization. Regrettably, seismic does not help in estimation of relative permeability and its spatial distribution in the field. Nonetheless, holistic comprehension of fluid production pattern, co-linking rock and fluid properties and other engineering and production data, underscores the importance

of an experienced and skilled seismic analyst for meaningful effective evaluation of 4D seismic for dynamic reservoir characterization. While fine-tuning quantification of porosity, permeability and saturation can be challenging task, 4D can be used as an indicator for other related applications.

Reservoir Drive Types, Water/gas Coning and Gas Phase Formation—4D Indicators

Reservoir Drive Types

Primary hydrocarbon recovery exploits the natural reservoir pressure (drive) for production which depends on the kind of reservoir, fluid type and saturation and the potential energy stored in them. The different type of reservoirs and drive mechanisms in oil and gas reservoirs can be enumerated as:

1. Gas cap drive – energy of compressed free gas cap overlying oil in reservoir
2. Water drive (aquifer) – energy from surrounding water in oil reservoir
3. Solution drive - energy of dissolved gas in oil reservoir, also known as depletion drive
4. Gravity drainage drive – energy due to gravity in an oil reservoir
5. Gas expansion drive – energy from compressed gas in gas reservoirs
6. Compaction drive – energy derived from compaction of reservoir rock.

The different drive mechanisms mentioned above are also tagged with type of reservoirs, type of fluid, oil and gas and their inherent natural source of energy (drive). Of the drive mechanisms mentioned above, the last named, the compaction drive mechanism is of different type and needs mention. The most common and powerful mechanism, the aquifer drive is also elaborated.

Compaction drive It is caused due to geomechanical effects on a reservoir under production and unlike other drives is not the energy inherent in the reservoir. As hydrocarbon is produced, reservoir pressure continues to deplete in the absence of external energy prop, and thereby increasing the effective pressure. This induces deformation in the reservoir and in clastic reservoirs it creates compaction of rocks which triggers an energy drive, discussed under 'Reservoir Geomechanics' later in the chapter.

Aquifer drive Aquifer driven reservoirs are fairly common and by far are most effective drive mechanism increasing recovery by 35–75% of initial oil in place. The mechanism can be bottom water or edge water drive (Refer Fig. 16). As oil is produced and the pressure drops, the surrounding water moves into the pay zone driving the oil upwards to the well bore and the process known as aquifer drive. In bottom water drive, the water underlying water from below moves up to drive oil, whereas in edge water the aquifer is located on the flanks and moves upward along the reservoir dip. In stratigraphic traps, however, the aquifer drive is edge water and from the single flank.

Appropriate drive mechanisms for optimal recovery from the reservoir are decided by the geological, petrophysical and reservoir engineering data, analysed by modelling and simulation studies. Nonetheless, clues from 4D seismic can be useful in supplementing and/or corroborating information on the drive active in the reservoir and possible immediate impending effects on production behaviour. Such imminent indicators of anomalous flow from 4D seismic may be detected which can also help in simulating other possible options as sensitivity studies for achieving the best recovery rate. 4D, calibrated with well and production data can detect and map the changed fluid contacts which can be used to re-estimate the remaining fluid volumes in the reservoir.

4D Indicators—Drive Mechanisms, Coning and Gas Phase Formation

Amplitudes are the simplest and straight forward means to monitor drive mechanisms and associated impending problems for concern. In favourable geologic situations, amplitude changes in 4D indicate the type of drive mechanism active in the reservoir and its fluid-flow efficacy. For instance, the 4D data may indicate uneven sweep of fluid from the reservoir due to

unforeseen reasons and can serve as precursors for likely remedial action to make alterations in production plans for improved reservoir management. In hydrocarbon reservoirs, particularly those manifesting seismic high amplitude anomalies in 3D, changes in amplitudes and size of anomalies can be symptomatic of drive types active in the reservoir and the linked fluid flow (Staples et al. 2006; Bousaka and O'Donovan 2000; Xu et al. 1997; He et al. 1997). The 4D amplitudes basically depend on depletion of pore pressure, type of fluid and change in phase saturation, i.e., the oil, gas and water mix (Anderson et al. 1997) The reservoir and production engineering data surely can show the current active drive and its effect around the wells but elsewhere, there can be isolated pockets, away from the wells which may be missed. Having mentioned about the various drive mechanisms in oil and gas reservoirs, their effect on changing the seismic amplitude anomalies in 4D, is briefly deliberated and summed up as below.

Gas-cap drive – Increase in size of bright amplitude anomaly in 4D, compared to 3D, can be indicative of expanding gas cap area, suggesting active gas-cap drive in oil reservoir.

Aquifer drive – Similarly, decrease in amplitude and size of the high amplitude anomaly in 4D may indicate oil depletion and water encroachment, signifying active aquifer drive.

Solution drive – Increase in amplitude can be suggestive of gas phase forming under solution drive (depletion) due to released gas from oil, a concern and caution for early remedial action.

Gas expansion drive – Sustained bright amplitudes in 4D without change with respect to baseline amplitude may imply de-pressurization of gas due to expansion drive in gas saturated reservoir.

4D seismic reservoir monitoring can also be useful in understanding anomalous fluid flow effects, known as 'coning' and discriminate the phenomenon as restricted locally to particular well(s) on localized scale or on a larger field scale. Coning poses production problems as bottom water from below moves upward or gas from gas-cap at top of oil reservoir infiltrates into the pay zone, choking oil production. The former is known as water coning and the latter gas coning and are usually considered local phenomena limited to particular well(s). Changes in size and strength of the amplitude anomalies in 4D, as discussed above, can be helpful in discriminating the anomalous fluid flow as localised or on field scale movement caused by aquifer drive/water flooding during secondary recovery and gas formation due to gas expansion. Similarly, primary production from oil reservoirs under natural depletion is driven by release of solution gas (solution drive) which after a period of time forms a gas phase that hampers production. Indication of such imminent problems indicated in time by 4D amplitude changes can help early redressals by appropriate action.

Amplitudes vary due to several reasons and the above discussions are limited to considering the changes conditioned only due to fluid content and fluid contacts under different reservoir drive mechanisms. Yet there can be ambiguities one may be careful of, for instance, while in one case dimming of amplitude indicates oil drainage under aquifer drive, in another case brightening of amplitude can be due to the oil-brine contact (Anderson et al. 1997).

Detection of Bypassed Hydrocarbon (Oil)

Detection of bypassed oil is an obvious corollary to what has been discussed above, that is, no amplitude change in 4D in a portion of reservoir under depletion would logically signify little or no production of hydrocarbon from that part. These are the interesting areas of unswept oil, known as bypassed oil that remains to be produced, and may warrant drilling additional in-fill wells. Further, production from a field results in decrease of saturation and drop in pore pressure ensuing water intrusion into the reservoir. Both the factors, the decrease in pore pressure and higher water saturation tend to raise the rock elasticity and consequently the seismic impedance. Relatively higher impedance of reservoir in 4D compared to the 3D baseline seismic, is therefore normally expected during production. On the other hand, areas indicating no change in impedances would suggest a status-quo, that is, no drop in saturation and pressure, which would

signify virgin reservoir status and thus identify the areas of untapped or by-passed oil. Yet another way to identify by-pass oil zones can be through pore pressure discussed earlier, where there would be no change in pressure shown in the base 3D The unswept oil under virgin reservoir pressure would not only lead to additional production but also for longer period and increase the field life. It is also interesting to know what caused the oil not to be swept, and left behind as bypassed oil? Impedances derived from 3 and 4D seismic inversion may provide the clues to connectivity problems in the reservoir impacting fluid movements.

Permeability Barriers and High Permeable Paths

Areas of by-passed oil are mainly created mainly by flow barriers such as shale layers and/or faults or highly permeable alternate paths segregating areas as compartments with no connectivity. Faults in particular depending on their kind can behave both ways as barriers as well as highly permeable paths for the 'flow units' within the reservoir to create unanticipated production behaviour. Faults with small displacements and fractures could occur during depletion or water injection process due to geomechanical changes (discussed later under 'reservoir geomechanics') and create anomalous fluid flow patterns. These subtle features can be detected and mapped from the 4D data through seismic attributes (Chapter "Analysing Seismic Attributes"). The fractures, developed anew or revitalized of existing fractures could act as barriers or provide highly permeable paths for the flow units in the reservoir which may result in uneven sweep of oil and show unanticipated early water and gas cuts at well heads. 4D amplitude time slices can detect the subtle faults in planview and can be mapped (Fig. 5) to understand their impact on connectivity of reservoir flow units and linked fluid flow.

Secondary and Enhanced Oil Recovery (EOR) Monitoring

After primary production for a period of time the reservoir loses its energy and requires artificial pressure to be induced, mostly by water or gas injection to improve hydrocarbon flow, the process known as secondary recovery. Enhanced oil recovery (EOR) processes, also increase productivity but unlike secondary recovery, involve reservoir stimulation which changes oil properties to improve flow. EOR methods are sometimes used as primary recovery process for producing hydrocarbon from unconventional reservoirs. Artificial stimulations such as water injection, gas injection and thermal flooding methods are commonly used to boost the depleting reservoir pressure or improving mobility of the fluid (heavy oil) for easy flow. In both the processes, success depends on the efficiency of the water/thermal sweeps which need to be monitored (Chapters "Borehole Seismic Techniques", "Oil Sands, Heavy Oil, Tight Oil/ Gas, BCGAs and CBM" and "Shale Oil and Gas, Oil Shale and Gas Hydrates").

Seismic 4D monitoring is primarily based on amplitude studies and is likely to be most effective in shallow young, unconsolidated sand reservoirs and under secondary recovery processes of water/ gas flooding. This is for the obvious reasons that such reservoirs are likely to show strong seismic amplitude and their changes can be easily noticeable in time lapse surveys (4D). In thermal methods, mostly used for heavy oils (Chapter "Oil Sands, Heavy Oil, Tight Oil/ Gas, BCGAs and CBM"), heat is applied to reduce oil viscosity for thinning the oil to flow and this lowers seismic properties that can be detected by 4D seismic. Though these are mostly applicable to sand reservoirs, successful monitoring of steam flooding by 4D in in-situ combustion for enhanced oil recovery is also reported from carbonate reservoirs (Xu et al. 1997).

Ideally, applications of 4D seismic can be utilized for a wide range of production related issues. However, having said that it may be emphasized that its efficacy essentially depends on the type of the problem, the data quality and more importantly on the imaginative, skilled and experienced analyst. Change in amplitude, as is well known, depends on several factors including the reservoir rock microstructure, the texture, pore geometry and nature of fluid and saturation. Change in 4D amplitude to be attributed to a

particular production related issue would depend on the expertise and experience of an analyst through synergistic studies of multidisciplinary data sets which could indeed be challenging.

Reservoir Geomechanics

Change in stress during depletion & secondary recovery/EOR processes creates strains in the reservoir rock and 'Reservoir Geomechanics' deals with such studies. The change in mechanical properties of rocks can cause unanticipated production problems and may partly contribute to divergences noticed during history matching of production behaviour with that planned. Deformations vary depending on the stiffness of reservoir rock, such as soft and hard and influence attendant fluid flow. Degree of deformation in the reservoir is also sensitive to several other factors including reservoir depth, geomechanical rock properties, fluid saturation, drive mechanism and reservoir pressure. Reservoir geomechanics, particularly, for big fields may require 4D seismic reservoir monitoring (SRM) to help address production related concerns.

Deformation and Linked Fluid Flow

Soft Reservoirs

Sands are relatively more compliant to stress and the strains are mostly grain- compaction due to increased stress (effective). Compaction increases density and causes reduction in porosity and shrinkage in pore throat that can lead to lower the relative permeability during production. This may be detected during reservoir monitoring by lower flow rates owing to reduced porosity & permeability. However, an interesting precursor to this can be the initial spurt in fluid production by induced compaction drive, a relatively short period phenomenon which could escape unnoticed. Conversely, during water injection, the reservoir pressure is relatively boosted (though not to the extent of original pressure) causing decrease in effective stress, which allows partial restoration of porosity and resultant increase in production. Particularly young unconsolidated offshore sands at shallow depths during water injection are likely to regain more of their original porosity as sands are known to be partially elastic.

Hard Reservoirs

Older sandstone and limestone reservoir rocks are relatively stiff rocks and under stress suffer structural deformations such as faults and fractures. The deformations may be reworking of existing and/or creating new fractures & fissures. Fractures create anisotropy in the reservoir that influences fluid flow patterns due to azimuthal dependent permeability. On the negative side, high permeable pathways created by faults can facilitate flow of water more than oil and the consequences can be unexpected chaotic flow including early water and gas cuts. This can also impair the planned water flood sweep and its efficiency resulting in unswept pockets of oil/gas left behind.

3D and 4D Seismic Roles in Exploration and Production

3D and 4D seismic technologies are proven cost-effective techniques though 3D is widely pervasive and the technique used extensively in petroleum exploration and production ventures. Time lapse 3D (4D), on the other hand are infrequent because of reasons cited below under subhead 'limitations'. The major benefits of 3D surveys can be summed up as below:

- Reliable delineation and characterization of structural and stratigraphic reservoirs.
- Providing input for reserve estimate and production profiles for field development.
- Reservoir monitoring of fluid movements and flow performances.
- Identification of by-passed hydrocarbon zones for in-fill wells to increase production.
- Detection of high-permeable paths and barriers to redress anomalous flow patterns.
- Decides optimal drilling locations for production, in-fill and injection wells.
- Assists optimal reservoir management.
- Minimises exploration risks, saving drilling costs of unproductive and dry wells.

4D Seismic: Limitations

4D seismic amplitude studies are the most convenient way of SRM but it may be underscored that it cannot be applicable to all fields. Mostly large fields with long life span may be the candidates for exercising 4D surveys. Its effectiveness in reservoir monitoring greatly depends on type of reservoir and the drive mechanisms. Seismic amplitudes for monitoring fluid flow requires DHI anomalies as a prerequisite and can be applied mostly in reservoirs which show high amplitude anomalies in 3D. Young oil and gas sand reservoirs at shallower depths are generally characterized by strong amplitude anomalies and under aquifer drive their changes in 4D seismic, particularly during primary recovery under aquifer drive are likely to be noticeable best for effective seismic reservoir monitoring (SRM).

Another problem of 4D is the stiff requirement to acquire identical data sets for baseline and monitor surveys. This is usually difficult to achieve as seismic acquisition and processing parameters are seldom same due to several technical, logistical and environmental problems. Even in marine 4D survey, the sea condition may be widely different than was during 3D acquisition which may affect data quality and analyses. Despite rigorous data conditioning for bringing the two sets of data to one working platform, it may still not be good enough to detect subtle changes in seismic response that can be attributed to only rock and fluid property or pressure changes in the reservoir. Geomechanical properties of reservoir rocks, elasticity, density, porosity, permeability, saturation and pore pressure changes occur during production impact seismic properties. However, these parameters varying differently may interact in a negative way resulting in the combined effect, that is too marginal, to create perceptible change in seismic response.

4D technique is complex, involving data processing and interpretation that require integrated workflow of multidisciplinary data. SRM is essentially an inverse problem that requires seismic inversions, seismic modelling and reservoir simulation. The 4D reservoir input parameters from seismic need to be authenticated by reservoir simulation and must be consistent with other information from multiple data sets like geological, well logs and engineering data. 4D needs infrastructure with fast interactive processing and, interpretation software capabilities with powerful visualization techniques that could be expensive and cost impediment and is commonly not a regular survey.

References

Anderson NR, Boulanger A, He W, Teng Y-C, Xu L, Meadow B, Neal R (1997) What is 4-D and how does it improve recovery efficiency? In World oil's 4-D seismic hand book, pp 9–13

Bousaka J, O'Donovan A (2000) Exposing the 4D seismic time-lapse signal imbedded the Foinaven active reservoir management project. OTC paper 12097

Brouwer F, Connolly D, Bruin GD, de Groot P (2008) Transformation and interpretation of seismic data in the Wheeler domain: principles and case study examples. Search and Discovery Article #40314

Chopra S, Huffman A (2006) Velocity determination for pore pressure prediction. CSEG Recorder 31, view issue

Etris EL, Crabtree NJ, Dewar J (2002) True depth conversion: more than a pretty picture. CSEG Recorder 26:1–19

He W, Anderson RN, Boulanger A, Teng Y-C, Xu L, Neal R, Meadow B (1997) Inversion of 4-D seismic changes to find bypassed pay. In: World Oil's 4D Seismic Hand Book, pp 29–32

Kolla V, Bourges Ph, Urruty JM, Safa P (2001) Evolution of deep-water tertiary sinuous channels offshore Angola (West Africa) and implications for reservoir architecture. AAPG Bull 85:1373–1404

Nanda NC (2018) Analysing a seismic pitfall – Pliocene superdeep high amplitude anomaly, offshore, Bay of Bengal, India. CSEG Recorder 43(7):30–33

Staples, R., J. Stammeijer, S. Jones. J. Brain, F. Smit and P. Hatchell, 2006, Time-lapse seismic monitoring-expanding applications: CSPG-CSEG-CWLS Convention, 181–189.

Xu L, Anderson RN, Boulanger A, Teng Y-C, He W, Neal R, Meadow B (1997) 4-D reservoir monitoring, the business driver. World Oil's 4-D Seismic Hand Book, pp 23–28

Zeng H (2006) Seismic imaging? Try stratal slicing. AAPG Explorer 28

Shear Wave Seismic, AVO and Vp/Vs Analysis

Abstract

Shear seismic data are extremely useful for determining rock-fluid properties. Nevertheless, shear wave properties and propagation mechanism are relatively complicated, acquisition of data (S-S) more expensive, and are often unviable in offshore. Land shear surveys are comparatively less frequent, and are occasionally acquired with OBC mode in offshore. Basics of shear wave properties, polarization, and mode conversion of P-waves to shear waves, their benefits and applications are described.

Fortuitously, conventional P-surveys generate mode-converted shear waves (P-Sv) that are recorded along P-reflections on land and in offshore. This offers excellent and convenient opportunity to analyze the combined P-and S-wave data. The benefits of joint studies provide more reliable prediction of rock-fluid properties and specifically help authenticate DHI anomalies for hydrocarbons through effective tools such as AVO and Poisson's ratio (Vp/Vs) analysis. Basics of AVO, types and attributes, impact of anisotropy, AVO modelling and Poisson's ratio and their applications are discussed in detail. Illustrations with seismic images, case examples, AVO limitations and pitfalls are highlighted.

'Birefringence', the unique S-wave splitting in anisotropic medium and used for fracture detection with its shortcomings are also included.

Compressional (P) wave reflection technology is the mainstay of petroleum exploration and continues to be so, to date, because the mechanics of P-waves are better understood and simpler to realize compared to that of transverse (shear) waves. P-waves with faster velocity, arrive earlier than S-waves, and more importantly, with much better signal-to-noise ratio. Moreover, compared to shear waves, compressional waves can be conveniently generated and recorded both on land and offshore and are easily adaptable for processing. This is why, despite shear waves proffering huge benefits in seismic analysis of rock-fluid properties, seismic technology applied in hydrocarbon exploration is mostly and commonly skewed in favor of P-seismic and so far been the focus of discussions in the earlier chapters.

The propagation mechanism for shear waves, on the other hand, is more complex. Shear waves are generated by a shear source and in horizontally layered isotropic medium, the S-wave movements are in two modes SV and SH with particle motions of SV in the incident vertical plane and that of SH in the orthogonal horizontal plane. Particle motions of both the S-waves are, however, perpendicular to direction of propagation. This is in contrast to that of P-waves propagating in isotropic medium where the particle motion is along the path of wave propagation. Another major difference with P-wave is

N. C. Nanda, *Seismic Data Interpretation and Evaluation for Hydrocarbon
Exploration and Production*, Advances in Oil and Gas Exploration & Production,
https://doi.org/10.1007/978-3-030-75301-6_9

that S-waves cannot travel in fluid medium. Shear wave surveys need horizontal source and detectors that are different from vertical detectors used in P-wave surveys. Shear wave reflection surveys (SS) on land are relatively more expensive and are not doable in marine environment due to unfeasibility. Shear waves are also difficult to be detected in low-velocity near surface layers and exhibit higher attenuation of reflected energy, poor signal to noise ratio and reduced resolution. Processing and interpretation of data are also relatively more complicated compared to that of P-waves. Nonetheless, the shear wave data, if made available, can be extremely useful for more accurate evaluation of seismic data.

Shear Wave Properties - Basics

In homogeneous isotropic media, relation between the P-wave velocity and elasticity is expressed by the equation

$$V_p = \sqrt{\frac{k + \frac{4}{3}\mu}{\rho}},$$

And for the S-velocity, it is

$$V_s = \sqrt{\frac{\mu}{\rho}},$$

where, ρ is the density of the rock, and the elastic moduli are the bulk modulus 'k' and the shear modulus 'μ'. As can be seen in the above equations, while the P-velocity depends on the bulk modulus, the shear modulus and the density, the S-velocity is dependent only on the shear modulus and the density. The shear wave velocity is much slower and as a rule of thumb, is considered about half of P-wave velocity.

An inherent property of shear wave is its inability to propagate in a fluid medium because fluids have little rigidity. Consequently, fluids in a rock hardly influence shear velocity (Vs), whereas they affect considerably the compressional velocity (Vp). The Vs for fluid saturated rocks may, however, exhibit marginal changes due to bulk density changes linked to fluid densities.

Polarization of Waves

Wave propagation in an elastic medium comprises of two components, the direction of wavefront motion, and the direction of motion of the medium particles. If during wave propagation the medium particles are displaced in a direction different from that of wave propagation, the waves are said to exhibit polarization. In S-waves, the particles of the medium are displaced perpendicular to the direction of wave propagation, and thus are inherently polarized. Depending on the direction of particle motion with respect to the direction of wave propagation, S-waves may be polarized in the vertical direction (SV-waves) or horizontal direction (SH-waves). On the other hand, for P-waves in a homogeneous, isotropic elastic medium, the wavefront motion and the motion of medium particles are along the same longitudinal direction and are nonpolarized. In an anisotropic medium, however, the polarization directions deviate from the direction of propagation and P-waves and S-waves are strictly neither fully longitudinal nor fully transversal (Kerner et al. 1989).

Shear wave while passing through an anisotropic medium such as fractured rocks shows another unique property – splitting into two orthogonally polarized waves. The two split wave's travel with different velocities and the phenomenon is known as '*birefringence*'. Birefringence and its utility in data evaluation is discussed later.

P-wave Polarization

The assumption of subsurface rocks providing a perfect elastic medium ignores the real earth situation, where seismic waves during propagation suffer absorption. Such energy-dissipating media are called anelastic, in which velocity of wave propagation is frequency dependent, and are also termed as dispersive media. In a non-dispersive media, all frequency components of a wave travel with the same velocity of propagation, and thus the wave is not polarized and a spherical wave front does not undergo change in shape as it travels. But in dispersive media, the different frequency components travel with different velocities, with the higher frequencies in the wave travelling faster than the lower frequencies resulting in change of wave front shape. The P-wave propagating in such dispersive media exhibits a 'wave-packet' propagation with changed wave front shape from an ideal spherical front. This results in P-wave exhibiting two velocities, the phase velocity and the group velocity (Fig. 1). The phase velocity is the velocity with which a particular wave phase travels and the group velocity at which the wave-packet propagates. The particle motions deviate from those in the direction of wave–packet propagation and the P-wave exhibits polarization. Group velocity is slower than the phase velocity which is usually considered as the seismic velocity.

Shear Wave Data Acquisition, Processing and Interpretation

Shear wave excitation on land is done by deploying a shear Vibroseis source which generates waves by oscillating horizontally. Shear waves can be excited by vertical vibrators also, but their recording requires large source-receiver offsets. In horizontally layered isotropic media,

the generation of shear waves, *SH* and *SV*, depends on the orientation of the source with respect to the receiver line. When the source is broad side to a receiver spread, *SH* waves are recorded with particle motions in X-Y plane. If the source is parallel to the detector spread, *SV* waves are recorded, the particle motions being in X-Z plane (Fig. 2). Both the *SH* and *SV* waves need horizontal geophones to be recorded. The propagating *SV* waves also have the property of creating P-waves at interfaces through 'mode conversion similar to inclined *P*-wave generating shear wave (*SV*), discussed later. *SH* waves, on the other hand, are polarized parallel to layered strata and provide relatively simpler records.

The deployment of Vibroseis on land, particularly in rugged terrains, can be logistically problematic and expensive. Further, appropriate horizontal geophones, different from the vertical geophones used for P-waves, are required to record the shear waves. Thus, for analysis of combined P-and S-wave data, two sets of survey equipment are required to record compressional reflection (*P-P*) and shear reflection (*S-S*) data which can be time consuming and expensive. Poor signal-to-noise ratio and low-velocity near surface statics, combined with problems of S-

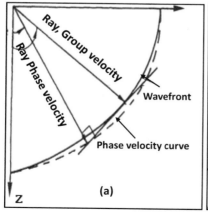 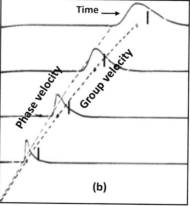

Fig. 1 Illustrating polarization of P-wave in dispersive media resulting in phase and group velocities. (**a**) Group velocity is at which the wave-packet (total energy) moves and is different from phase velocity which is the velocity at which a particular phase of the wave travels. Note the change in shape of spherical wave front (phase velocity curve) to group velocity wave front. The respective ray paths normal to the wave fronts are shown (**b**) Phase velocity is always higher than the group velocity. (after Anstey 1977)

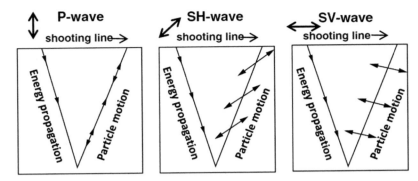

Fig. 2 Sketches illustrating the source-receiver geometry for recording (**a**) compressional (P) wave, (**b**) shear (SH) and (**c**) shear (SV) waves. Particle motions for P-and S-waves are also denoted. P-wave source and receivers are in vertical plane (**a**), which is in contrast to shear wave recording in horizontal plane (**b**) and (**c**). SH is recorded by the broad side receiver while SV by the in line receivers with respect to source. SH and SV wave recording requires horizontal geophones. (After Ensley 1984)

wave splitting in anisotropic medium, make processing of *S*-wave data more complicated. Moreover, interpretation of both P-and S-data jointly can also be worrisome. Due to large differences in propagation velocity (arrival time of an event) and in wave form characters of P-and S-waves, it becomes difficult to identify and correlate the P-P and S-S reflections from an interface, time-wise or character-wise. (Fig. 3). Without a satisfactory correlation, a joint analysis of *P*- and *S*-amplitudes and velocities for more reliable interpretation becomes meaningless. Due to these drawbacks in acquisition, processing and interpretation (API) of data, shear-wave surveys are rather infrequent.

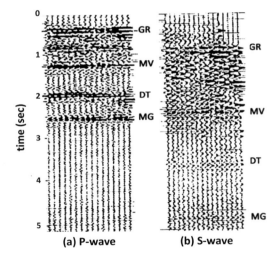

Fig. 3 Seismic images of (**a**) *P*-and (**b**) *S*-wave reflections. *S*-reflections (**b**) show more noise and less continuity compared to *P*-reflections (**a**). Note the nonlinear delay in *S*-arrival times of the geologic units with respect to P-times. Dissimilarity in reflection characters and differences in arrival times of *P*- and *S*-reflections cause difficulties in correlation of horizons creating problems for combined analysis. (Modified after Robertson and Pritchett 1985)

Mode Converted Shear Waves, P-SV, P-SH and P-SV-P Waves

As mentioned earlier, when a P-wave strikes a reflecting boundary at oblique incidence, in addition to reflected and transmitted P-waves, it generates reflected and transmitted S-waves known as mode converted waves. For normal incident P-waves in isotropic media, however, there is no mode conversion. The mode converted waves are the *SV* component of shear waves and can be recorded on the surface by horizontal geophones (*P-SV*). It can also be recorded by vertical geophones (*P-SV-P*) as *SV*

waves have the property of getting reconverted to P-waves at boundaries. Mode conversion occurs every time an inclined P-wave encounters an interface and the ensuing *SV* wave which in turn gets reconverted to P-waves on encountering a rock interface. This is how the mode converted

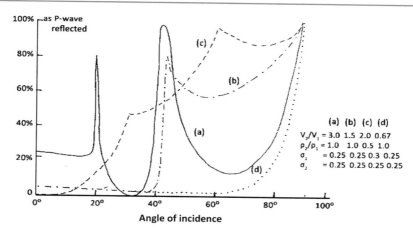

Fig. 4 Illustrating reflected P-amplitude patterns with increasing incidence angle for varying contrasts of velocity, density and Poisson's ratio indicated by models **a, b, c** and **d**. In curve (**b**) for velocity contrasts common in exploration, note the amplitude maxima at critical angle due total internal reflection. With no density and Poisson's ratio contrast, the amplitude change can be solely attributed to the velocity contrast. Curve (**a**) having similar parameters but with very high velocity contrast shows two amplitude maxima and illustrates the nonlinear behavior of amplitude variation with velocity contrast. With density and Poisson's ration factored in, the curve (**c**), shows amplitude patterns getting complicated. Also notice the remarkable curve (**d**), parameters similar to that of curve (**b**) but with small negative velocity contrast. It continues to show weak amplitudes without peaking to maxima as it does not reach critical angle (after Sheriff (1975), Courtesy, EAGE)

shear waves (*P-SV-P*) are eventually recorded along with P-wave reflections. The degree of mode conversion is mainly dependent on the angle of incidence and the velocity contrast and on anisotropy in case of propagation in anisotropic medium. The balance between reflected and transmitted P- and S-energy, changes with angle of incidence. For small angles of P-incidence, most of the energy is carried by the P-reflected and P-transmitted waves whereas with increase in the angle of incidence, mode conversion becomes more prominent. This aspect of converted shear wave energy being strong in far offsets is amply utilized in predicting rock-fluid properties. The P-amplitude variation with incidence angle is a non-monotonic complex function governed by Zoeppritz's equation. Applications of mode converted shear wave are discussed in detail later under AVO analysis.

Sheriff (1975) has shown the variation of computed P-amplitude curves with angle of incidence for models with varying contrasts of velocity, density and Poisson's ratio between two adjacent media (Fig. 4). The graph provides

interesting and valuable insight to patterns of angle dependent amplitudes. Curve (b) in the figure, the model with the velocity contrast, which is commonly met in exploration, shows reflected amplitude diminishing slowly till the maxima reached at the critical angle ($\sim 40°$), where 'total internal reflection' occurs. Since there is no density and Poisson's ratio contrast factored in the model, the amplitude change can be attributed solely to the velocity contrast. Intriguingly, the curve (a) for the model with similar parameters but with much higher velocity contrast, however, shows two amplitude maxima and illustrates the nonlinear behavior of amplitude variation with high degree of velocity contrast. When density and Poisson's ratio contrasts are factored into the model, the amplitude patterns in the curve (c) get complicated. And the curve (d), is interesting; with the model having the same parameters as in curve (b) but with small negative velocity contrast where the amplitude response continues to show low values and with no maxima, because negative contrast does not reach critical angle to cause total internal reflection. It may be noted that curve (a) with very high velocity contrast (three times) is significant for exploration of sub-basalt and sub-salt prospects, where two notches of amplitude maxima occur with little energy reflected within the intervening zone between the two incident angles ($\sim 20°$ and $\sim 40°$). Curve (c), on the other hand, shows that wide-angle reflections corresponding to sub-basalt and pre-salt formations immediately underlying high-velocity formations such as Mesozoic limestones/sandstones, yielding negative contrast, may show low amplitudes and mired in noise and may be imperceptible in seismic.

In on land geologic basins where older rocks are likely to have large velocity contrasts, mode conversion can be expected to be more pronounced. However, the mode conversion reflection geometry involves shift in conversion point (CP) from that of P-wave reflection mid-point (MP) and poses processing complications. The shift in CP added by the problem of asymmetrical geometry of the P-incident and the converted S-reflection make converted wave data processing more complicated (Fig. 5). Processing of converted wave data is a separate operation and requires special efforts other than for P-P processing. Most processing centres may not

have the converted wave processing capability and the upshot is the regular processed P-data contains converted waves as noise. An interface with very high impedance contrast like that of an intrusive body within a sedimentary section can generate strong mode-converted waves which may interfere with primaries and impact resolution of in conventional P-wave data.

P-seismic surveys allow natural generation and recording of shear waves, on land as well as in offshore because of mode conversion. P-SH and P-SV-waves are indeed a welcome spin-off of P-surveys though they have to be recorded with horizontal geophones and require special processing. However, the P-SV component, reconverted to P-wave (P-SV-P) can be recorded by conventional vertical geophones both onland and offshore. In conventional marine surveys, the first mode conversion occurs when the P-wave gets converted to S-wave at the sea bottom as this offers the first velocity contrast boundary. The mode converted SV wave travelling downwards gets reflected at interfaces and travels upwards till the sea bottom where the wave gets mode-converted to P-wave (P-SV-P) and recorded by the conventional streamer geophones (Fig. 6). Mode conversion is likely to be most efficient at the sea bottom when the velocity contrast is large with comparatively hard rocks comprising the sea bottom. In contrast, in offshore deep water areas where the sea bottom is usually soft and loose, the mode conversion efficiency may be hampered. Incidentally, shear information is engrained in P-seismic as a byproduct and without additional cost and efforts has been extremely useful for combined analysis of P- and S-wave attributes and has found much favor with explorationists.

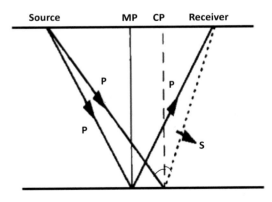

Fig. 5 Schematic showing geometry for recording of mode converted (P-S) waves on land. Note the asymmetrical ray trajectory for the reflecting S-wave causing shift in CDP midpoint and in the mode converted conversion point (CP) which requires separate and extra processing efforts than the conventional P-P processing. (After Stewart et al. 1999)

Multicomponent Surveys (3C and 4C)

Multicomponent seismic surveys on land record P- and S-waves (P, SH, SV) by deploying P- and S-sources and three orthogonally oriented geophones, two horizontal and one vertical and is known as 3C survey. In offshore areas, with conventional air gun as P-source, P-waves and

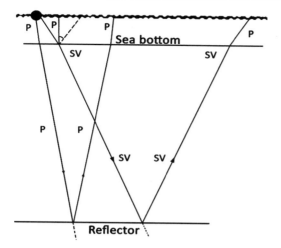

Fig. 6 Schematic illustrating recording of converted shear waves (*P-SV-P*) in offshore. *P*-wave is converted to shear (*SV*) wave at the sea bottom and after getting reflected from an interface, travels upward and gets mode converted to *P*-wave at the sea bottom (*P-SV-P*) and is recorded by conventional streamer geophones along with *P*-wave reflections. (After Tatham and Stoffa 1976)

mode converted *S*-waves are recorded (*P*, *P-SH and P-SV*) by three component geophones planted on the sea bottom. A streamer hydrophone is also added as a fourth component to record conventional *P*-waves and the survey is referred to as *4C* acquisition. *P*-waves are detected by the Z-component geophone and the hydrophone while S-wave components, the X and Y are recorded by the horizontal geophones. Because the geophones and the connecting cables are laid on the ocean bottom, it is known as ocean bottom cables (*OBC*) acquisition. Multi-component *P-S* recording, however, is relatively expensive, especially in offshore environment. Also formidable problems like asymmetric binning, shear statics and shear wave splitting due to azimuthal anisotropy make data processing more complex for most regular data processing centers.

Benefits of Shear Wave Studies

Shear wave applications have many-fold benefits. P-wave reflection data alone may not be able to predict rock-fluid properties conclusively as rocks with varied properties may show near-

similar P-response. Since the *S*-wave characteristics and their seismic responses vary differently from *P*-waves, it provides supplementary means to assess and validate interpretation of *P*-data.

Imaging Below Offshore Gas Chimneys

In offshore, presence of near-surface gas chimneys and diffusions, may cloud *P*-imaging of an underlying reservoir while the *S*-waves, being insensitive to fluid saturation, can effectively image the reservoir. (Fig. 7). Also, consider cases, where *P*-impedance contrast for an interface is too small to cause a reflection, where *S*-impedance contrast may be adequate to form a perceptible image. Reservoir properties are mostly analyzed from the three essential seismic properties, the amplitude, the velocity and the impedance of the *P*- and *S*-waves and ideally, both the *P*- and *S*-data, wherever possible, need to be jointly studied for better prediction of rock-fluid properties. Shear wave studies though are known useful in a wide range of geologic applications, a few common but important applications are described below.

Validating DHI (P-wave Amplitude Anomalies)

Bright spots are amplitude anomalies associated with hydrocarbon reservoirs, more commonly with gas sands and are considered as direct hydrocarbon indicators (Chap. "Direct HydrocarPbon Indicators (DHI)"). However, many bright spots, on drilling, were found to be devoid of hydrocarbons, which necessitated validation of bright spots prior to drilling to avoid dry wells. This is accomplished by joint study of *P*- and *S*-wave data. The concept has been there since early years, where the far offsets of conventional *P*-records, containing the converted *S*-wave energy were picked and separately processed to provide the supplementary *S*-stack amplitude section (Tatham and Stoffa 1976). Once the *P*-and *S*-stack sections are created from the prestack gathers of recorded *P*-data, amplitude of the DHI anomaly can be easily compared for

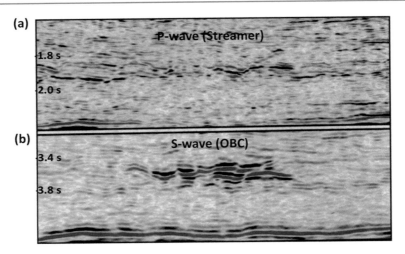

Fig. 7 Seismic image showing (**a**) *P*-(normal streamer) and (**b**) *S*-wave ocean bottom cable (OBC) seismic recording in offshore highlighting the benefit of S-wave. The geologic features, imaged clearly (**b**) by shear waves in central and bottom part in OBC is not imaged by P-wave (**a**) in conventional streamer mode. (After Stewart et al. 1999)

authentication. This is illustrated by a geologic model of a gas sand with water contact and its computed *P*-and *S*-seismic amplitude responses (Fig. 8). The discrimination is based on the fact that while gas saturation considerably impacts *P*-amplitude it does not affect the *S*-amplitude. Accordingly, bright amplitudes seen in *P*-stack (Fig. 8b) but absent in the *S*-stack (Fig. 8c), would qualify the anomaly as hydrocarbon bearing, whereas, presence of amplitude anomalies in both sections would indicate the amplitude anomaly due to lithology. Similarly, a dim-spot anomaly in *P*-data with no such corresponding anomaly in *S*-data can be suggestive of the presence of gas in a carbonate reservoir. Analysis of *P*- and *S*-velocities carried out for authenticating hydrocarbon bearing anomalies are discussed in detail later. However, extraction of shear wave information from normal P-data for building an *S*-stack section at the processing center can be painstaking and cumbersome and highly subjective depending on data quality, survey geometry and analyst's bias. The methodology is quite similar to comparing the near sack and far stack seismic amplitudes, as is discussed in Chap. " Direct Hydrocarbon Indicators (DHI)" which involves mode converted shear waves at far offsets due to large angle of P-incidence. However, with

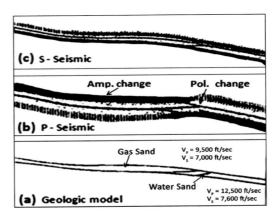

Fig. 8 Schematic shows seismic modelling of *P*-and *S*-wave for validating *P*-amplitude 'Bright spot 'gas-sand. (**a**) geologic model, (**b**) computed *P*-seismic and (**c**) computed *S*-seismic. Note the high amplitude reflection of the gas sand and change of polarity (indicated by arrow) at gas–water contact in P-seismic (**b**) while the reflection amplitudes are weaker and without any change in polarity in *S*-seismic (**c**), as shear waves are insensitive to fluid. (Modified after Ensley 1984)W

advent of innovated techniques, a more robust approach utilizing *P*- and S-amplitudes from trend analysis of amplitude variations with offset from P-wave pre-stack data, known as AVO, has become a standard current industry practice.

AVO Analysis for Validation of Hydrocarbon Sands

AVO study deals with two complimentary attributes, the change in amplitude with offset across a NMO corrected prestack gather and its gradient, the measure of the amplitude change. Analyzed together, it delivers more dependable predictions.

AVO Amplitude Analysis

Though AVO is primarily a study of variation of *P*-amplitudes with offset, it is included in the chapter for shear wave, as AVO phenomenon involves shear waves. The reflectivity of a *P*-wave varies with angle of incidence (offset) because of mode conversion. While.at normal incidence, the reflectivity is known as P-normal reflectivity, for inclined incidence it is called the *P*-angular reflectivity. A normal move out (NMO) corrected prestack gather which includes all the recorded traces from a depth point, shows the amplitude variations with offset signifying the *P*-angular reflectivity. While amplitudes at small angles of incidence (near offset) is dominated by *P*-impedance contrast, at larger angles (far offsets), due to mode converted shear waves, it is controlled by contrasts in Poisson's ratio (σ), often expressed simply by the ratio of *Vp/Vs*, discussed more in detail later in the chapter. The *P*-amplitude variation with (offset) due to mode conversion is thus influenced by set of contrasts in each of these properties, the *Vp*, the *Vs* and the density of rocks across the interface. The relationship is mathematically expressed by a set of complex Zoeppritz's equations, which provide a linkage between the variations of *P*-reflection coefficient with angle of incidence (θ) for a given primary and shear velocity. The *Vp/Vs* ratio is the dominant parameter that controls the variations in *P*-amplitude patterns with angle of incidence, which varies depending rock and fluid type. For hydrocarbon saturated rocks the *Vp/Vs* is generally lower.

AVO Classification

AVO is mostly applicable to siliclastic sand hydrocarbon reservoirs and more often in gas saturated sands where the effect of amplitude anomaly is greatly pronounced. The geologic models to compute seismic response are usually defined by a set of values of *Vp*, *Vs*, and ρ for the hydrocarbon saturated sand and the overlying shale/tight rock acting as seal. Depending on the variations in contrast of these properties at the interface, AVO anomalies for the reflection from the gas sand top are categorized as Class 1, 2, 3 and 4. Though the classifications are widely recognized for gas sands, oil saturation, too, can show similar AVO behavior, may be with relatively reduced effect. AVO amplitude effect is best seen for gas-sands capped by shale, as 'bright spot' anomalies and is the reason for the gas-sand models accepted universally for the AVO classification. Amplitude variation with offset for a gas sand is illustrated schematically in a plot of reflectivity versus angle of incidence with AVO classifications annotated (Fig. 9). The figure also shows intercepts and gradients for classes of AVO, an attribute discussed later in this chapter. The cause of the AVO amplitude patterns and their types are explicated below.

Class 1 AVO (High-Impedance Gas Sand) Model

The class 1 AVO anomalies are caused by high impedance gas sand, well compacted and relatively less porous reservoirs with large compressional velocity and density (high impedance) with shale as cap. The gas-sand and shale interface presents positive contrast for *Vp* and ρ (density) and negative Poisson's ratio, 'σ', ($\sim Vp/Vs$). Referring to Fig. 9, the amplitude shows a maximum positive value at normal incidence after which it decreases to zero at the crossing of x-axis at a fairly large angle of incidence. Beyond the angle, the amplitudes again increase but with polarity reversed, becoming negative. The plot of reflectivity with incidence angle also shows a negative gradient. Gradient denotes the way the amplitude changes, described later.

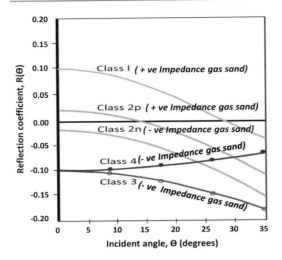

Fig. 9 Graphic illustrates the four classes of AVO anomalies for gas sands showing *P*-amplitude variation with increasing angle of incidence (offset). The characteristic properties of the gas sands causing the AVO anomalies are annotated. Normal reflection coefficient of reflections from sand top and its polarity signage are seen on 'Y' axis. The curves also give an idea about the way the amplitude change with angle of incidence (the gradient)

The AVO pattern can be seen in the modelled responses for high impedance sandstone saturated with gas and with water along with the actual observed seismic (Fig. 10). In the prestack synthetic gather for the model, the high amplitude at near offset is seen to decay with increasing offsets till it becomes too feeble (Fig. 10a). The reversal to negative polarity however is not clear, perhaps due to inadequate long offset considered in the model. This corroborates well the decaying pattern seen in the actual seismic response (Fig. 10b). As is to be expected, no variation in amplitude with offset (AVO effect) is noticed in the modeled response for the sand saturated with water. (Fig. 10c).

The reversal to negative polarity, however, occurs at a fairly large offset; an offset seldom used in routine acquisition and thus may be missed. Too large offsets also carry associated propagation problems and noise that can obscure the reflection quality and its polarity. However, in a conventional normal stack section, class 1 AVOs may be difficult to observe as near and far

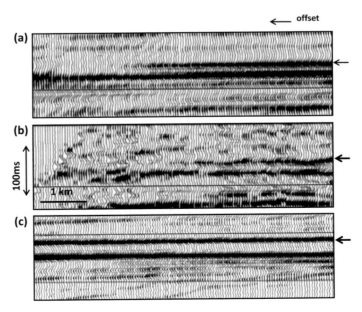

Fig. 10 Figure shows (**a**) computed amplitude prestack gather of a model for Class 1 AVO (high impedance gas-sand), (**b**) real seismic gather and (**c**) synthetic gather for water saturated sand. The arrow shows the positive reflectiin amplitude from top of gas-sand diminishing with increasing offset (angle of incidence) in (**a**) and confirms the seismic response (**b**). No amplitude change is seen for the modelled water sand (**c**). (After Downton 2005)

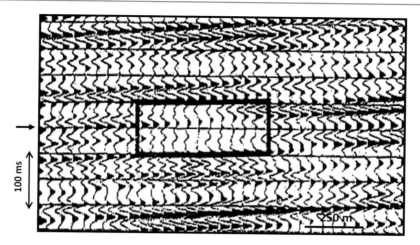

Fig. 11 Seismic segment showing Class 1 AVO anomaly which may not be perceptible in normal full-angle stacks due to weak amplitudes . Positive amplitudes in near traces due to high-impedance gas- sand, decrease with offset and reverses polarity at large offsets to negative . Since all traces in a CDP are stacked with both near-positive and far-negative amplitudes, the reflection amplitude (arrow marked) is compromised and imperceptible. (After Rutherford and Williams 1989)

traces with opposite polarities are summed up over the entire spread of gather, which could result in weak amplitude to be noticed (Fig. 11). This is interesting and noteworthy, because deeper and older, gas saturated sandstone with no noticeable amplitude on normal stack sections could have been exploration targets that are missed. Also it may be underscored that such high impedance hydrocarbon reservoirs of older age are likely to show dim-spots and high amplitude anomalies linked to Pre-Eocene age, *prima-facie*, may be normally dubious.

Class 3 AVO (Low-Impedance Gas Sand) Model

The class 3 AVO anomalies, in contrast to Class 1, are caused by low impedance gas sands, usually of relatively young age, and more explicitly at moderate depths. They usually have a large negative normal-incidence reflection coefficient, which becomes more negative as offset increases. The anomalies appear on stack sections as very strong amplitude anomalies and are known as the 'bright spots' (see Chap. "Direct Hydrocarbon Indicators (DHI)"). The interface between gas-sand and the overlying shale is characterized by a large negative P-impedance contrast with velocity, density and the Vp/Vs ($\sim \sigma$) of gas-sand

being appreciably lower than those of the overlying shale. Mio-Pliocene and younger age unconsolidated gas-sands generally typify the model all over the world. The amplitude variation with offset, computed for a real model is shown (Fig. 12). The gather shows reflection with high amplitude and negative polarity from the top gas sand in the near traces which continues to increase with angle of incidence (offset). The figure also illustrates the AVO pattern for the reflection from the base of sand which shows strong positive reflectivity increasing with offset. The plot of reflectivity with angle also indicates a negative gradient, similar to Class 1 (refer Fig. 9). An example of bright spot seismic anomaly as seen in near and far-angle stacks and the pattern of negative amplitude increasing with offset across the prestack gather typifying class 3 AVO is shown in Fig. 13.

Class 2 AVO (Near-Zero Impedance Gas Sand) Model

Moderately compacted gas-sands and having impedance close to overlying shale exhibit Class 2 AVO anomalies. The normal impedance contrasts are small and can be either positive or negative depending on the properties of the sand and the overlying shale. In case of positive

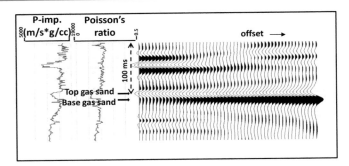

Fig. 12 Figure shows computed amplitudes for an actual model of low impedance gas sand with negative Poisson's ratio contrast. The gather shows Class 3 AVO anomaly with negative amplitude (trough) increasing (more negative) with offset for the sand top reflection. The sand bottom reflection shows the positive amplitude (peak) becoming more positive with offset. (Image courtesy: Arcis Seismic Solutions, TGS, Calgary)

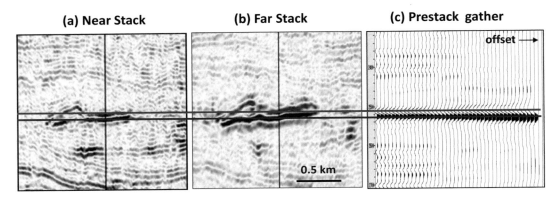

Fig. 13 Showing seismic example of 'bright spot' anomaly seen in near (**a**) and far angle stacks (**b**). The prestack gather (**c**) shows Class 3 AVO validating the interpretation of low impedance gas sand. Note the top and bottom reflections of the gas sand with strong negative and positive amplitudes increasing with offset, becoming stronger (after Nanda 2017)

reflectivity, amplitude is small at near offset and decreases with offset and are called Class 2p (positive) anomalies (Fig. 9). Class-2p AVO anomalies are similar to Class-1 AVO but with polarity reversal occurring at small incidence and are likely to be missed in normal stack sections though strong amplitudes can be expected at adequately large offsets but with negative amplitude. Class 2n (negative) AVOs, on the other hand, though have small amplitude at near trace, it increases with offset and can be considerable in far offsets to be noticed. For this reason, Class 2n AVO anomalies may be easier to detect by a simple comparison of angle-stacks amplitudes (Chap. "Direct Hydrocarbon Indicators (DHI)"). Referring to Fig. 9 again, the reflectivity versus

angle plot for Class 2 AVO shows negative gradients similar to Class 1 and class 3. Hydrocarbon sands of Class 2 AVO type with marginal impedance contrasts can be illusory and may or may not always correspond to amplitude anomalies on stacked data (Rutherford and Williams 1989). However, an offshore Pliocene hydrocarbon sand showing Class 2n (negative) AVO with poor to no amplitude in near but strong amplitude in far offset stack is exemplified (Fig. 14).

Class 4 AVO (Low-Impedance Gas Sand with Hard Top Seal) Model

Class 4 anomaly occurs where low impedance gas-sand reservoir is capped by a comparatively hard rock, such as calcareous shale or limestone.

(a) Near stack   **(b) Far stack**   **(c) AVO gather, class 2n**

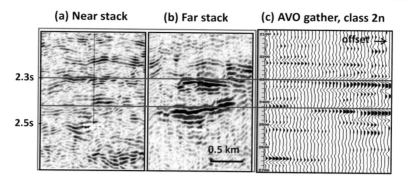

Fig. 14 Example of seismic image of two offshore gas sands one above the other in (**a**) near- (**b**) far-angle stacks and the prestack gather (**c**) which shows AVO anomaly of Class 2n. The near stack shows poor/no amplitudes for the sands (marked by lines), whereas the far stack (**b**) shows strong amplitudes. This can also be seen in the prestack gather (**c**) where the poor amplitudes in near trace increase with offset showing strong amplitudes. The mismatch between the strengths of amplitudes in seismic image and the stack gather is due to amplitude bias used differently. (after Nanda and Wason 2013)

Impedance and *Vp/Vs* contrast of the gas reservoir being lower, it creates a strong negative contrast for the reflection from the top of reservoir. However, the amplitude decreases with increasing offsets, becoming less negative. This is in stark contrast to amplitude changes seen in Class 3 AVO where the negative amplitudes become more negative with offset. Despite the strong negative P-impedance for the gas-sand, similar to class 3 type, the AVO pattern is opposite, because the gas-sand shows positive *Vp/Vs* in contrast to that in class 3 and other classes of AVO where the Vp/*Vs* contrast of gas sand is negative with respect to cap rock. This is unique and can happen when the *Vs* (∼ shear impedance) of the hard cap rock is high which lowers *Vp/Vs* resulting in the gas-sand to have positive *Vp/Vs* contrast. The *Vs* value, thus, may have to be considered as a discrete input in modeling to compute AVO response. This aspect is discussed later under Poisson's ratio. The seismic response of class 4 AVO shows a large negative reflectivity at zero offset with very high amplitude, diminishing with increasing offset and has singularly different, the positive gradient. An onland seismic example of a Mesozoic sand showing Class 4 AVO anomaly with gradient is shown in Fig. 15.

Gradient and Intercept Analysis

AVO analysis in pre-stack gather is the simplest and most convenient way to look for the presence of hydrocarbon, using the criteria of increase and decrease in amplitude with offset. However, amplitudes can be corrupted due to several reasons and affect AVO amplitude analysis. For instance, the trace amplitudes improperly weighed and balanced and uncertainty in picking the exact polarity of reflectivity (positive/ negative *Rc*) especially for thin hydrocarbon reservoirs due to noise and limited resolution could hamper AVO results. In some cases, the polarity convention displayed in seismic may be unknown to the interpreter as it varies from company to company.

Another quick way for AVO studies is by using the gradient of a plot of amplitude against square of sine of angle of incidence. Amplitude variation in the prestack gather and the intercept-gradient plots are complimentary and comprise the two attributes of AVO analysis. Shuey (1985)'s two term approximation of Zoeppritz's equations, for small incidence angles (up to ∼ 30°) gives the simple linear equation below, satisfying a straight line.

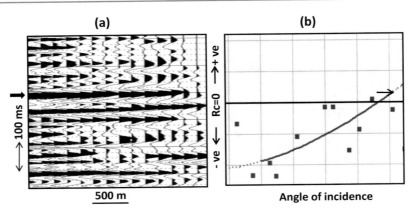

(a)   **(b)**

500 m   Angle of incidence

Fig. 15 Seismic example of Class 4 AVO anomaly. (**a**) prestack gather showing high amplitude, negative polarity (peak, black) reflection from top gas sand decreasing with offset, (**b**) plot of reflectivity versus incident angle showing the negative normal reflectivity and the positive gradient, characteristics of Class 4 AVO (Image: courtesy ONGC, India)

$$R(\varphi) = A + B\sin^2(\varphi),$$

The equation expresses a convenient straight line relationship between P-reflection coefficient $R(\varphi)$, the intercept (A), and the gradient (B) of amplitude changes when plotted against square of (sine of) angle of incident (φ). This is illustrated for a Class 3 AVO showing amplitude variations in the prestack and the plots of reflectivity against incidence angle as well as the square of sine of incidence angle (Fig. 16). Plot of angle of incidence versus reflectivity shows a curve (Fig. 16b) whereas the plot versus square of sine of angle a straight line (Fig. 16c). The best straight line fit to the plot of reflection coefficient against square of sine of angle provides the gradient (B) and the intercept (A) on the y-axis at zero offset. The intercept 'A' (positive/negative) shows the reflectivity at normal incidence (Ro) with polarity and the AVO gradient 'B' (positive/negative), indicates the rate of amplitude change with offset, the angular P-reflectivity (Ostrander 1985; Castagna et al. 1998).

Displaying the product attribute 'AxB' highlights an anomaly that can be interpreted as a gas/oil indicator characterizing DHI amplitude anomalies, particularly belonging to Class 3 AVO type showing a positive 'AxB'. However, the product indicator does not work well for Class 2 anomalies and is hard to interpret without prior knowledge of the type of gas sand reservoir in the area. The Class 1 and Class 4 AVO anomalies also cannot be discriminated on this basis of product attribute as both are negative. The AVO attributes, the amplitude and the intercept (A) and gradient (B) product for the three major types may be summed up as below.

(a) Class 1, 2p ... A +ve, amplitude decrease with offset, *Gradient −ve*, Product AxB − ve.
(b) Class 3, 2n ... A −ve, amplitude increase with offset, Gradient −ve, Product AxB + ve.
(c) Class 4 ... A −ve, amplitude decrease with offset, *Gradient + ve*, Product AxB −ve.

Cross-Plot of Intercept and Gradient

Castagna et al. (1998) has suggested classification of AVO anomalies based on plot of intercept and the gradient (Fig. 17b). The 'A' (normal reflectivity, Ro) and 'B' (the gradient), which controls the angle-varying reflectivity can be computed for a model using well data parameters to establish the background trend of shales, and water saturated sands, denoted by the trend line passing through the centre (Fig. 17b). Deviation in trends observed in cross plots generated from seismic from the base line at the centre can be a good indicator of gas/oil sands (Fig. 17a). Normal reflectivity and gradient for prestack CDP gathers in the zone of interest can be derived from the seismic data volume in the workstation

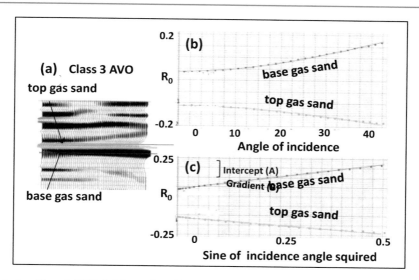

Fig. 16 Illustrating Class 3 AVO (bright spot) attributes for gas sand. (**a**) Gather shows strong amplitudes with negative and positive polarity for the top and bottom sand reflections respectively, increasing with offset. (**b**) Plot of reflectivity versus incidence angle showing a curve fit whereas (**c**) plot of reflectivity versus square of sine of incidence angle shows a straight line. The straight line provides negative gradient and its intercept at 'Y' axis shows the normal reflectivity with its polarity signage of the reflection from top of gas sand. Bottom reflection characteristics are also shown (Image courtesy, Satinder Chopra, Calgary)

in an efficient and conveniently quick way and cross plotted to indicate potential hydrocarbon sands. AVO anomalies for reflections from top of hydrocarbon sands fall below the central line and for the sand-bottom, above it (Castagna et al. 1998). The indications can then be validated by AVO gather amplitude analysis. However, modeling requires a priori knowledge of the rock and fluid parameters in the area, desirably from well data. A quantitative AVO analysis, more sophisticated than the routine study of AVO intercept ('*A*') and gradient ('*B*') attributes, is to estimate *P*- and *S*-impedance from seismic inversion for predicting reservoir properties. The technique is known as AVO inversion and is outlined in Chap. "Analysing Seismic Attributes" under seismic forward modelling and inversion.

Poisson's Ratio, Angular Reflectivity in AVO Analysis

Castagna et al. (1998) have shown the normal incident reflectivity '*A*' is determined not exactly by the contrasts in velocities and densities at an interface but by their normalized contrasts, expressed by the equation

$$A = 1/2(\Delta Vp/Vp + \Delta\rho/\rho),$$

where, $\Delta Vp = Vp_2 - Vp_1$, $\Delta\rho = \rho_2 - \rho_1$ are the velocity and density contrasts in the two media, $Vp = \frac{1}{2}(Vp_1 + Vp_2)$ and $\rho = \frac{1}{2}(\rho_1 + \rho_2)$ are the average of velocities and densities. Thus, for a given impedance contrast, the reflectivity is likely to be higher in smaller ranges of *Vp* at shallower depths compared to that at deeper depths. The term '*B*', the gradient (slope), on the other hand, involves additional rock parameters, the shear velocity *Vs* and the Poisson's ratio ($\sim Vp/Vs$) and is controlled by the parameters the ΔVp, *Vp*, $\Delta\rho$, ρ and ΔVs, *Vs* and Poisson's ratio (Castagna et al. 1998), where ΔVs denotes the contrast in shear velocity in the two media and Vs, the average shear velocity ($Vs_1 + Vs_2/2$). While the normal reflectivity (*Ro*) is influenced by normal P-impedance contrast which determines the polarity and amplitude at zero offset, the change in P-amplitude with offset (P-angular

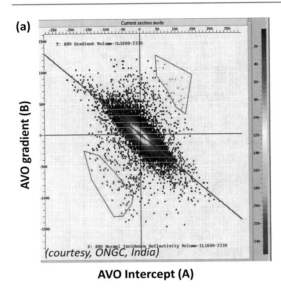

(a)

AVO Intercept (A)

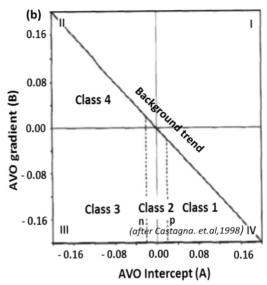

(b)

AVO Intercept (A)

Fig. 17 Shows cross plot of AVO intercept (A) versus gradient (B) for identifying hydrocarbon sands. (**a**) Cross plot generated from offshore seismic data and (**b**) model computed by Castagna, et al. The central line passing through centre (**a** and **b**) denotes the back ground trend corresponding to water sands and shales. Deviations from the trend indicates gas sands. Note in (**a**) the clusters below and above the line indicating gas sand tops and bottoms. The graphic (**b**) illustrates the characteristic AVO gradients for all classes of AVO (Courtesy, (**a**) ONGC, India and (**b**) after Castagna et al. 1998)

reflectivity) and its measure of change, the gradient (G), is significantly influenced by the elastic modulus, the Poisson's ratio ($\sim Vp/Vs$).

The AVO anomaly patterns involving the increase and decrease of amplitude and their degree of change (gradient) with angle of incidence in a prestack gather are thus essentially determined by the P-impedances and Poisson's ratio contrasts. Knowing these two contrasts and the way they vary can then predict the AVO attributes. This also holds true for the four classes of AVO anomalies for hydrocarbon sands discussed earlier and can be generalized as a rule

- If both the impedance and Poisson's ratio contrasts change the same way the amplitude increases with offset and
- Conversely, if the changes are in opposite way, the amplitude decreases.

This is exemplified by computed synthetic models (Fig. 18). When contrasts change the same way, the peak or trough, controlled by impedance contrast, increases with increasing offset, i.e., positive amplitude becomes more positive (Fig. 18a) and negative becomes more negative. The Class 3 AVO for bright spot belongs to this type where negative amplitude becomes more negative with increasing offset. Conversely, when the contrasts change in opposite way, the peak/trough depending on impedance contrast, decrease with offset; i.e., positive amplitude becomes less positive and negative amplitude becomes less negative (Fig. 18b). The Class 1 and Class 4 AVO belong to the former and the latter pattern.

Representative rock properties, AVO types with their attributes, the amplitude, intercept, gradients and their product is summarized (Table 1).

Rock properties vary greatly depending on depositional environments and consequently can show many patterns of amplitude variation. Variogram plots of computed reflection coefficients for several varying rock parameters are demonstrated by Ostrander (1985) and

(a)

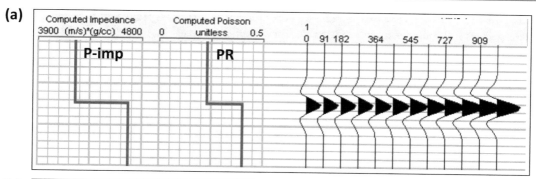

(b)

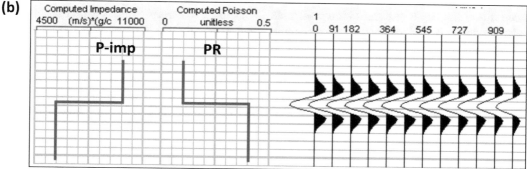

Fig. 18 Illustrating the role of *P*-impedance and Poisson's ratio contrasts in determining the amplitude variation pattern in AVO prestack gathers, (**a**) Both contrasts varying in same way, amplitudes increases with offset., positive becomes (in this case negative amplitudes) more positive and likewise, negative amplitude become more negative as in Class 3 AVO. (**b**) If contrasts vary the opposite way, amplitudes decrease with offset, negative (in this case) becomes less negative and positive becomes less positive as in Class 1 AVO (image courtesy, Satinder Chopra, Calgary)

Table 1 Representative rock and seismic properties showing AVO types and attributes for gas sands

AVO types (gas sands)	Class 1 AVO		Class 3 AVO		Class 4 AVO	
Top seal lithology	Shale	Gas sand	Shale	Gas sand	Cal.sh/Lst	Gas sand
Vp (m/s)	2520	3010	2620	2540	2940	2590
Vs (m/s)	1310	1750	1320	1560	1820	1490
ρ (g/cm^3)	2.4	2.2	2.3	2.1	2.3	2.2
P_{imp} (m/s. g/cm)^3	**6050**	**6620**	6030	5330	6990	**5700**
Vp/Vs (PR)	1.9	1.7	2.0	1.6	1.6	1.7
AVO attributes	R_0 (A) + ve, PR − ve, Contrasts vary opposite way, +ve Amp. decrease with offset, Gr(B) − ve Product AxB − ve		R_0 (A) − ve, PR − ve, Contrasts vary same way, − ve Amp. increase with offset, Gr(B) − ve Product AxB + ve		R_0 (A) − ve, PR + ve, Contrast vary opposite way, − ve Amp. decrease with offset, Gr(B) + ve Product AxB − ve	

Table summarises the rock and seismic properties of gas sands that cause AVO types and attributes. Essentially, the impedance and Poisson's ratio contrast determine the AVO amplitude pattern; if the two contrasts vary opposite way, amplitude decreases with offset (Class 1 and 4 AVO) and vary the same way amplitude increases with offset (Class 3 AVO). The class 4 AVO occurs when low impedance gas sand is capped by a tight rock having *Vs* higher than *Vs* of the gas sand leading to lower Poisson's ratio (PR) for cap rock

Castagna et al. (1998). However, modelling pre-stack gathers factoring the six sets of variables for the two media can illustrate the AVO anomalies better. In offshore settings, the exploration targets are often the channel cut and fill and fan complexes where sand reservoirs capped by shales mark the hydrocarbon habitats. The channel-cut and fan complex facies commonly vary laterally and considerably showing different rock properties. The properties of the cap shale can also vary appreciably depending on clay content, mineral composition and depth of burial. With such wide variances in rock-fluid parameters of sands and the cap shales, the amplitude and its manner of change with offset across the gather can be of myriad types. To give an idea, examples of synthetic elastic gathers computed for a shallow Pliocene mudrock section with varying

parameters (actuals from log) in a deep offshore well, East coast, India is shown in Fig. 19.

In panels (Fig. 19a, b), the P-impedance and the Poisson's ratio contrast for the second medium is positive and shows positive amplitude (peak) increase across the gathers but with marginal difference, the gradient (order of amplitude change) being a tad slower. This is due to the density contrasts being opposite which demonstrates the role of the density in AVO response. Considering the bottom panels (Fig. 19c, d), the impedance contrasts in both panels are negative and extremely small, whereas the Poisson' contrasts are positive but considerably larger in panel c). The AVO responses though are similar in pattern, it is significantly strong in c). The miniscule negative amplitudes at the near traces decrease with offset to become zero and further

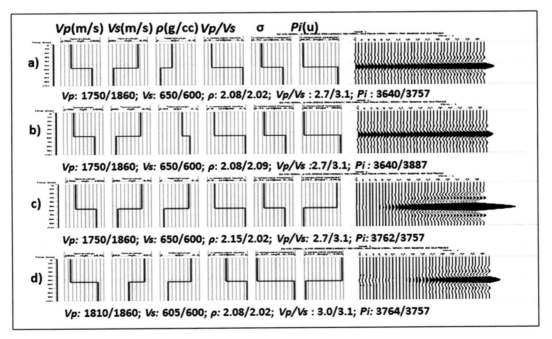

Fig. 19 Computed elastic gathers to show patterns of amplitude variation with offsets for different parameters of an offshore mudrock formation. In panels (**a**) and (**b**), the P-impedance and the Poisson's ratio contrast for the second medium are positive and show positive amplitude (peak) increase with offset. The tad slower gradient in (**b**) is attributed to density contrast being opposite. In panels (**c**) and (**d**), the impedance contrasts are minimal

and negative, while the Poisson' contrasts are positive, strikingly large in panel (**c**). The AVO responses are similar but significantly stronger in (**c**) due to large PR contrast. Note the miniscule negative amplitudes at the near traces decrease with offset and reverse polarity to positive and the rapidly increase with offset showing very strong amplitudes with steeper gradient (**c**). (courtesy, Satinder Chopra, Calgary)

reduce by reversing the polarity. The positive amplitude from thereon continues to increase rapidly with offset, flaring to stronger amplitudes and steeper gradient in c), demonstrating that stronger the positive amplitude larger is the decrease. This is attributed to the high Poisson's contrast and clearly shows its dominance in controlling the AVO amplitudes in far offsets. However, it is mathematical as in real cases of long offsets such amplitude patterns are unlikely because the wave energy would die down. Ostrander (1984) had also demonstrated the impact of Poisson's ratio on amplitude which gets more pronounced with larger contrast.

It is also interesting to note the prestack gathers in the upper panel exhibits amplitude increase with offset and with AxB product positive similar to 'bright spot' anomaly showing Class 3 AVO (Fig. 19a, b). This, however, is a false indication and can be verified by the positive signage of polarity; Class 3 AVO characteristically shows negative polarity ($-ve$ Rc). Castagna et al. (1998) had cautioned of such false AVO anomalies arising from nonhydrocarbon related reflections based solely on positive AxB product which can be due to the high Vp/Vs values often occurring in very soft shallow brine saturated sediments. A real case showing the apparent 'bright spot' seismic anomaly and false AVO response arising due to strong Poisson's ratio contrast is exemplified.

Case Example

The case example is from offshore deep waters, East Coast, India, where large contrast in Poisson's ratio within a shallow thick mudrock section caused high amplitude events in 3D seismic (Fig. 20a). The well logs show the thick Pliocene section of unconsolidated, brine saturated mudrocks having large contrasts in Poisson's ratio but meagre contrasts in impedance in the upper part of the section (Fig. 20b) which is correlated with the high amplitude events on seismic (Fig. 20a). The computed AVO response modeled with primary and shear velocities V_P and V_S, the density ρ and Vp/Vs values, picked

from the logs, shows the positive amplitude increasing rapidly to be significantly strong at the far-offset traces in the gather (Fig. 20c). This looks similar to Class 3 AVO but is false because of the wrong polarity, the positive Rc. The model also clearly demonstrates the dominance of Poisson's ratio in influencing angular reflectivity with normal P-impedance playing a secondary role (Nanda 2017).

Anisotropy Effect in AVO Analysis

Anisotropy, particularly the intrinsic anisotropy (VTI) type in shale which caps the gas sands can be a major aberrant in AVO analysis. Direction dependent seismic velocities may lead to imperfect NMO corrections, impacting trace amplitudes in the prestack gather for AVO analysis. Consequently, estimates of AVO attributes, the amplitude, the gradient (B) and the intercept (A) which is utilized for estimation of normal reflection coefficient (Ro) will suffer if not compensated for the anisotropic effects. The anisotropic impact can be significant, particularly in cases where a thick shale (VTI) unit overlies the reservoir.

Wright (1987) in his seminal paper has shown computed curves for P-reflection coefficient as a function of incidence angle for sand models (1) overlain by isotropic shale and (2) by anisotropic shale (Fig. 21). The model parameters V_{11}, V_{33}, V_{55}, V_{13} in the figure are independent elastic constants of a transverse isotropic media explained in detail in his paper. Without going into the intricacies and for easy understanding it may be simply stated that the V_{11} and V_{55} parameters mentioned above relate to Vp and Vs. For isotropic shale (model 1), the positive amplitude increases with incidence angle whereas, for anisotropic shale (model 2), it decreases. The Poisson's ratio of both type of shales is lower resulting in positive PR contrast for the sand. However, the P-impedance contrast of the sand is positive with respect to the isotropic shale (model 1) whereas, it is negative for the anisotropic shale (model 2). Consequently, the PR and Impedance contrasts varying in same

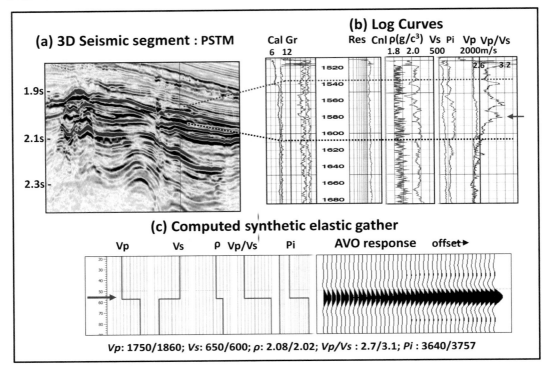

(a) 3D Seismic segment : PSTM

(b) Log Curves

(c) Computed synthetic elastic gather

Vp: 1750/1860; Vs: 650/600; ρ: 2.08/2.02; Vp/Vs : 2.7/3.1; Pi : 3640/3757

Fig. 20 Composite figure showing (**a**) high amplitude offshore shallow seismic anomaly (**b**) log curves showing mudrock section with large contrast in *Vp/Vs* in the upper part (**c**) computed AVO gather for the high amplitude anomaly. The high amplitude anomaly is correlated to high *Vp/Vs* contrast in the logs (marked) with no significant *P*-impedance contrast. The synthetic gather (c) shows near trace positive amplitude increasing considerably at far offsets due to positive contrasts in impedance and *Vp/Vs*. Note the amplitude pattern apparently looking like a genuine AVO anomaly for the nonhydrocarbon mudrock which actually is false. (After Nanda 2017)

way in model 1, the positive amplitude increases with angle of incidence. For model 2, however, the impedance and Poisson's ration contrasts vary the opposite way and the positive amplitude decreases with angle of incidence. It is interesting to note that the normal reflection coefficient for model 2 which is positive becomes negative at larger angles of incidence when velocity of the anisotropic shale gets higher than sand velocity and the positive amplitude starts decreasing.

Wright's observation on anisotropy has great significance in exploration. Anisotropy effect in case of Class-3 AVO anomalies underlain by anisotropic shale are therefore likely to show higher amplitudes at longer offsets. Consequently, water-saturated young-age soft sands

with velocity close to shale, which often occur in deep offshore environment, may show weak positive amplitude in near-offset but reverse polarity and increase with angle of incidence showing strong amplitudes at far offsets. This can be mistaken with Class 2n AVO anomaly and may falsely indicate hydrocarbon saturation. Kim et al. (1993) have also stated about amplitude increase with offset for bright spot anomalies (Class 3) getting somewhat more pronounced due to anisotropy (VTI).

Anisotropic shale may also affect AVO attribute analysis by blurring 'AxB' cross plots which determine the background shale trend to help identify the hydrocarbon sands. Particularly, thick young-age shales in marine setting having

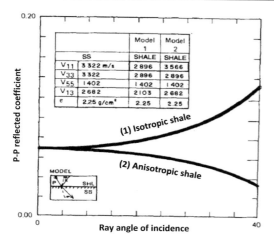

		Model 1	Model 2
	SS	SHALE	SHALE
V_{11}	3 322 m/s	2 896	3 566
V_{33}	3 322	2 896	2 896
V_{55}	1 402	1 402	1 402
V_{13}	2 682	2 103	2 682
e	2.25 g/cm³	2.25	2.25

Fig. 21 Showing computed curves of P-reflection coefficient as a function of angle of incidence for sand model overlain by (1) isotropic shale and (2) anisotropic shale. The parameters V_{11} and V_{55}, denote the Vp and Vs of sand, isotropic and anisotropic shale, shown in the table. The Poisson's ratio of both type of shales is lower resulting in positive PR contrast for the sand. For isotropic shale (model 1), the P-impedance is positive and so positive amplitude increases with incidence angle whereas, for anisotropic shale (model 2), it becomes negative and amplitude decreases with offset. Note the near trace positive impedance contrast becomes negative as the velocity of the anisotropic shale gets higher than sand velocity at increasing incidence angle. (after Wright 1987)

lateral changes in properties may show heavy scatters in the cross-plot with no set central trend and affect AVO analysis. Shale anisotropy effect may be manifested in seismic during data processing by the 'hockey stick' appearance of a corrected NMO gather (Fig. 22). The up-turned shape of the gather at the far end is somewhat akin to an overcorrected gather with lower NMO velocity whereas in the far traces the wave has actually travelled faster with higher velocities due to the anisotropic shale. Similar up-turn 'hockey stick' effects seen in the reflection events below would confirm the anisotropic phenomenon and can be corrected by way of residual velocity corrections, However, correction due to anisotropy requires anisotropic coefficients which are often not obtainable and can be cumbersome task and the issue mostly remains unattended.

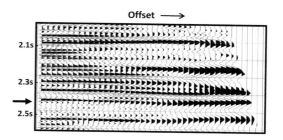

Fig. 22 Seismic manifestation of anisotropy in NMO corrected gather from deep offshore 3D data. Note the *'hockey-stick'* look akin to an overcorrected NMOs at 2450 ms (arrow marked) suggesting anisotropy (*VTI*) in the shale above. Overcorrection is due to lower NMO velocity applied instead of the actual higher velocity due to anisotropy through which wave travelled. Similar effect seen also for the event below supports inference of anisotropy in the overburden. (*Image courtesy: ONGC, India*)

AVO Modelling, Elastic and Zoeppritz's Gather

Several factors can affect the amplitude of reflection in a prestack gather and can be e constraints in proper analysis of AVO (Allen and Peddy 1993; Ostrander 1985). Acquisition and processing deficiencies in seismic data can cause spurious AVO effects and it is important the seismic input is examined for data quality prior to AVO analysis so that the data can be cleaned and improved by reprocessing/reconditioning. Wave equation modelling of elastic and Zoeppritz gathers computed and compared with real seismic allows review of quality of existing data for reliable AVO analysis and interpretation (Fig. 23). While the elastic gather computation includes primary, multiples, mode converted waves and diffractions (Fig. 23a), the Zoeppritz gather computes only the primary reflections events (Fig. 23c). Comparison of the computed model responses with real seismic can reveal the inadequacies, if any, in the seismic data (Fig. 23b). For instance, the primary reflection event seen near 1.6 s in elastic and Zoeppritz gathers (Fig. 23a, c) is absent in seismic (Fig. 23b). More significantly, the reflection at 1.7 s, prominent in Zoeppritz but somewhat subdued in the elastic gather due to interference

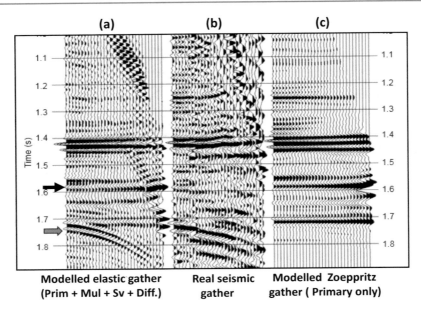

(a)   **(b)**   **(c)**

Modelled elastic gather   Real seismic   Modelled Zoeppritz
(Prim + Mul + Sv + Diff.)   gather   gather (Primary only)

Fig. 23 Figure showing modelled elastic gathers (**a**), real seismic gather (**b**) and the modelled Zoeppritz gather (**c**) to assess seismic data quality before attempting AVO analysis. Elastic gather (**a**) includes the primaries, multiples, converted waves (*Sv*) and diffractions and the Zoeppritz gather (**c**) only the primary events. Note the primary reflection event seen at 1.6 s in (**c**) and in (**a**) is absent in seismic (**b**). More significantly the reflection event at 1.7 s in Zoeppritz model which indicates positive AVO anomaly for hydrocarbon (C) has feeble indication in elastic model due to diffraction (**a**) and absent in seismic (**b**). It underscore the necessity for reconditioning the seismic data (after Chopra and Marfurt 2012)

of diffractions is completely obliterated in seismic due to diffraction and other noise. This is an important reflection event suggesting a potential hydrocarbon bearing target and missing it in seismic could be a lost opportunity. The Zoeppritz modelled gather involving primary reflections is ideally the obvious desired seismic and needs reconditioning of data.

Limitations of AVO Analysis

Despite detailed appraisal of AVO, qualified as a strong hydrocarbon indicator, many wells drilled for oil/gas are found dry. This suggests shortcomings in AVO technique which the interpreter needs to be careful. Firstly, the Zoeppritz's equations, which allow estimating the reflection coefficient, the mainstay of the technique, is valid for a single interface only and with several other constraints. Assumptions of plane waves instead of spherical and non-inclusion of layering effects are some factors that may affect adversely the computation of reflection coefficients in model response needed for calibrating field seismic (Allen and Peddy 1993). Referring back to Fig. 4, the P-amplitude variations with angle of incidence as computed by Sheriff (1975) is non-monotonous and can be highly complicated. Thin-bed effects (tuning thickness), composite events from overlapping of reflections, restriction to $\sim 30°$ incident angles (approximations to Zoeppritz's equations used for AVO analysis), presence of multiples and noise, heterogeneity and anisotropy in the reservoir and overburden, scattering, absorption and dispersion effects in an otherwise ideally assumed isotropic homogeneous media for wave propagation are several factors that may affect the results of AVO analysis.

In addition to the above constraints, data acquisition and processing deficiencies may be one of the major factor to cause spurious AVO effects impeding meaningful study. The feasibility study of data for a proper AVO analysis through AVO modelling, discussed earlier, requiring reprocessing for conditioning of the data is the most important step to improve signal-to-noise ratio. This includes removing acquisition footprints without compromising resolution; stringent amplitude balancing, and efficient pre-stack migration to provide true reflectivities. Careful reprocessing of seismic data results in enhancement of effective signal bandwidth with noise muted, preserves relative true amplitudes and offers better resolution of images. Although these are standard measures adopted in routine 3D processing, the data may yet lack the desired quality. Unfortunately, it may not be always achievable due to several reasons, one of which may be the interpreter has no time or access for reprocessing.

Though AVO anomalies are categorized in four types for hydrocarbon sands, it is possible to have deviations in AVO attribute patterns as rock-fluid parameters of reservoir and of the overlying rocks vary greatly in diversified geologic settings. Rock and fluid properties are also dependent on temperature, pressure, degree of diagenesis and compaction, mineral composition and texture, saturation and viscosity, etc. These factors can significantly affect AVO analysis results and assigning a classification does not guarantee the presence of a commercial prospect, especially the gas.

Yet another innate limitation of AVO technique is that it works well mostly in siliclastic reservoirs and often in a particular depth range, known as 'AVO Window'. The window is usually relatively moderate, where the nature of the geologic setting and the seismic data are favorable for such amplitude studies. It may be recalled that at greater depths, the pore-fluid effect on seismic is less perceptible (Chap. " Seismic Wave and Rock-Fluid Properties") and the image quality may be obscured due to reduced resolution and signal-to-noise ratio caused by absorption loss, presence of multiples and other wave propagation effects, particularly in onland data. This again emphasizes the point that AVO modeling, desirably precede AVO

analysis for reliable prediction of hydrocarbon sands.

Vp/Vs Analysis, Prediction of Rock-Fluid Properties

Seismic amplitudes though are straight forward and convenient for prediction of rock and fluid properties, they are also impacted by several factors other than geologic and predictions can be uncertain on stand-alone basis. Velocity analysis, on the other hand, is more dependable for the purpose though a conjoined study is likely to yield better results. Sedimentary rocks are defined by matrix, porosity and fluid contents, which affect P- and S-wave velocities differently. The P-velocity is generally more sensitive and is commonly used for predicting reservoir properties. But predictions based on changes in P-velocity alone can sometimes be ambiguous as it is also impacted by factors such as lithology, pore fluid, clay contents and fractures. Further, several rocks may have overlapping P-velocities making it difficult to establish the causal relation. The S-wave velocity, on the other hand, behaves in a different way due to the dissimilar wave properties and become useful in assisting reliable identification of rock-fluid properties. The varying differential responses of the P- and S-velocities to elastic moduli of rocks can be normalized by their ratio, the Vp/Vs, which is a robust indicator of rock and fluid properties than any single velocity analysis can alone achieve. The Vp/Vs is a ratio directly related to Poisson's ratio 'σ', which is considered an important lithology and fluid discriminant (Chap. "Seismic Wave and Rock-Fluid Properties"). Poisson's ratios for sedimentary rocks are reported to commonly vary between 0.2 and 0.4. A hard rock such as limestone has the highest value of Poisson's ratio, around 0.38, followed by shales and water saturated sandstones. Poisson's ratio for gas saturated sands mostly exhibit a much lower value of 0.15–0.17, widely considered as a crucial index for gas saturation. Prediction of rock and fluid properties from *Vp/Vs* analysis is discussed below with a case example.

Vp/Vs for Fluid Prediction

It is common to express the Poisson's ratio by *Vp/Vs* as the velocities are conveniently measured in well and estimated from seismic inversion (see Chap. "Seismic Modelling and Inversion"). Compressibility of fluid in a rock greatly affects *P*-wave velocity but hardly influences *S*-velocity as fluids have negligible rigidity. While, presence of gas, a highly compressible fluid, in pore space lowers *P*-velocity considerably, water-saturated rocks tend to show higher *P*-velocity, as water is relatively less compressible. *S*-velocity, on the other hand, is seldom affected in either case, except for marginal effects due to change in fluid density. Since reservoir fluid lowers its bulk modulus (*k*) noticeably but marginally affects the shear modulus, the *Vp/Vs* acts as an excellent indicator of pore fluid. Lower values of *Vp/Vs* in general indicate hydrocarbon saturated rocks whereas higher values indicate water saturated sands, shales and nonreservoir rocks. The *Vp/Vs* values are typically lower, 1.5–1.8 and the shales and water sands higher with 2.2–2.4, the lower values mostly considered in the industry as standard index as gas/oil indicators. However, the *P*- and *S*-velocities of rocks, besides fluids, can be significantly controlled by texture, porosity and pore shape of rocks deposited under diverse depositional and tectonic styles (Nanda 2017). Consequently, the *Vp/Vs* can exhibit values over a wide range and may result in ambiguous interpretation and is illustrated by an offshore case example.

Case Example

Pliocene hydrocarbon bearing channel/fan complex sands, East coast, India show excellent DHI anomalies. The sands are typically assortment of channel, point bar, levee and crevasse splay facies that vary significantly affecting the amplitude and the *Vp/Vs* values. Several high

amplitude DHI anomalies on drilling encountered oil-sands in some wells, gas-sands in others and water sands in a few. The hydrocarbon sands show wide range of *Vp/Vs* values varying from 1.8 to 2.4, measured in sonic logs (Fig. 24). A study of several wells in the area reveals good and shaley oil-sands show *Vp/Vs* values of 1.8 and 2.1 to 2.2 respectively (Fig. 24b, a). The shaley gas-sands, however, in some instances, show unusual high values in the range of 2.4, an index generally assigned to shales and water-sands (Fig. 24c). The high *Vp/Vs* value is found to be due to relatively more lowering of shear modulus due to shaliness than of the bulk modulus of the sand. However, the most significant and interesting observation is the good watersand that shows strangely low *Vp/Vs* value of 1.8 (Fig. 24d). This is because of the higher *Vs*, presumably due to rock minerology which resulted in lowering of *Vp/Vs*, a value considered widely an index for oil-saturated good sands.

The example highlights the point that in some geologic situations matrix can play more dominant role than the fluid in influencing the *Vp/Vs* (Poisson's ratio) and cause large variance in values leading to overlapping. The *Vp/Vs* variances are more likely in offshore channel sand and fan complexes where the rock-fluid properties of soft shallow siliciclastic reservoirs and of the top shale cap vary greatly. In such scenarios, reliable prediction of type of sands and its fluid contents from seismic may become problematic as seismic images provide combined response of all the rock constituent elements. Under the circumstances delineating and characterizing the reservoir sands only by *Vp/Vs* values derived from seismic simulation inversion techniques may not be fruitful. On the other hand, combined analysis of angle stack amplitudes (refer Chap. "Direct Hydrocarbon Indicators (DHI)") and AVO attribute can to a large extent be useful to provide better solutions. The example also shows that the *Vp/Vs* may not be always the dependable index for predicting fluid. Ccommonly recognized values of *Vp/Vs* of 1.5–1.7 for gas and 1.6–1.8 for good oil sands and the higher values (2.0–2.4) for water sands and/or shales are empirical and typically are area-specific. The

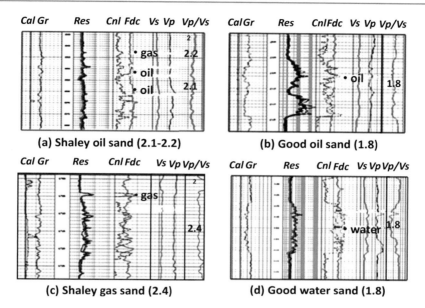

(a) Shaley oil sand (2.1-2.2)

(b) Good oil sand (1.8)

(c) Shaley gas sand (2.4)

(d) Good water sand (1.8)

Fig. 24 Panel displaying wide range of *Vp/Vs* values varying from 1.8 to 2.4 for hydrocarbon bearing sands depending on type of matrix and fluid. (**a**) Shaley oil sand shows Vp/Vs ∼ 2.1, (**b**) good oil sand shows Vp/Vs ∼ 1.8, (**c**) shaley gas sand shows unusually high Vp/Vs ∼ 2.4 and (**d**) good water sand shows inexplicably low Vp/Vs ∼ 1.8. Such large variations and overlapping of *Vp/Vs* makes it hard to predict fluid content reliably at times. (after Nanda 2017)

Vp/Vs values may be taken only as a guide to evaluate DHI anomalies for hydrocarbon in the area under exploration subject to well log and bench-marking the *Vp/Vs* indices.

Vp/Vs for Lithology Prediction

The *Vp/Vs* ratio is also sensitive to lithology and can especially help where P-velocity is within a range of overlapping values for more than one type of rock lithology. The lithologies can be discriminated by *Vp/Vs* as the *S*-velocity varies differently from P-velocity for rocks. For instance, low *P*-velocity can be attributed to shales and unconsolidated sands but can be discriminated by *Vp/Vs* which would be lower than that of shale. Similarly, Oligocene sandstones and Miocene limestones may show near similar *P*-velocities in the range of 3200 m/s and difficult to infer the correct lithology. But with *S*-velocity of sandstone usually being higher than that of limestone, a lower *Vp/Vs* of around 1.6–1.7 for sand would discriminate it from limestone with a value of 1.8–2.0 (Wang 2001; Pickett 1963; Castagna et al. 1998). Another useful application of *Vp/Vs* can be in estimating the sand-shale ratio, considered an important geologic factor for petroleum exploration. Shaliness decreases *Vs* to a large extent with relatively less impact on *Vp* of sand, resulting in large values of *Vp/Vs* for highly shaley sands. Increasing *Vp/Vs* would show more shaliness whereas decreasing values indicate highly sandy reservoir indicating sand-shale ratio. McCormack et al. (1984), in a case study has computed for varying sand-shale ratio in a model of 400 ft formation illustrating the responses (Fig. 25).

In exploration for carbonate reservoirs, *Vp/Vs* of tight limestones can discriminate it from porous carbonate rocks by the former's higher values (Wang 2001). Similarly, limestones and porous dolomites, may exhibiting similar *P*-impedances, but can be discriminated based on higher *Vp/Vs* value of limestones and help map the carbonate reservoir facies (Rafavich et al.

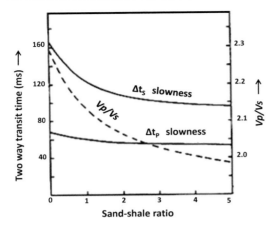

Fig. 25 The plot illustrating variations of Vp(slowness, Δt_P), Vs (slowness, Δt_s), and Vp/Vs with change in shale content computed for a model of 400 ft clastic formation. Increase in shaliness (decreasing sand-shale ratio) reduces considerably the Vs (increase in Δt_s) but to a lesser degree the Vp, resulting in higher Vp/Vs. (After McCormack et al. 1984)

1984). This could greatly help in exploration for reservoirs in wide spread limestone platforms which often are tight formations with discretely developed porosity pods. A scan of Vp/Vs attribute volume obtained from 3D seismic inversion would enable to find the porous patches. Shale plays key roles in petroleum prospecting for both conventional and unconventional reservoirs, depending on their properties linked to depositional environment which can at times be interpreted from Vp/Vs analysis. Clay-rich shale with relatively less rigidity, are likely to show higher Vp/Vs values, of the order 2.4 and more and can indicate deep water marine depositional environment (Nanda and Wason 2013).

Vp/Vs for Porosity and Clay Content Prediction

Predicting porosity and clay content in a reservoir are crucial to reservoir characterization. Increased porosity and clay content both decrease P- and S-velocity of a sand reservoir due to lowering of the bulk modulus and the rigidity, though with varying order. While the P-velocity lowering is affected more significantly

by pore geometry than the porosity, per se, the S-velocity is affected to lesser extent by porosity. Consequently, Vp/Vs may not be a reliable indicator for porosity. However, increasing clay content in matrix is expected to show higher Vp/Vs values due to considerable decrease in rigidity compared to bulk modulus. Specifically for offshore shallow exploration targets like Pliocene and Pleistocene hydrocarbon saturated channel sands, Vp/Vs derived from seismic inversion may be effective in differentiating good clean channel sands from clayey levee' sands, though the fluid prediction may be ambiguous, discussed earlier in the case example (refer Fig. 24).

However, it may be stressed that the Vp/Vs values indicated for lithology and fluid saturation are empirical and only denote order and range of values for in typed geologic area. The caution is because the Vp/Vs responds to both matrix and fluid which include texture, pore geometry, shaliness and fluid saturation. The dominant factor controlling the seismic response therefore must be clearly established from well log data and rock physics modelling for reliable prediction of rock and fluid properties. Petrophysical analysis is crucial for the interpreter to fully comprehend seismic rock physics.

Shear Wave Birefringence, Fracture Prediction

Shear wave travelling in anisotropic medium (naturally fractured reservoir) splits into two orthogonally polarized waves which propagate with different velocities. The Fig. 26 illustrates a shear wave entering an anisotropic medium composed of vertical fractures with polarization oblique to fracture plane oriented NW-SE. The wave splits into two orthogonally polarized waves, S_1 (NW-SE) and S_2 (NE-SW) and the splitting of wave phenomenon is known as *birefringence*. The polarized wave S_1 travelling parallel to fracture plane is faster than the polarized wave S_2 travelling across and the difference in velocities can be measured in seismic by their arrival times. In reservoir with

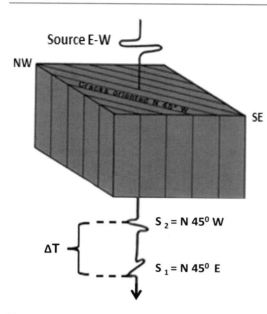

Fig. 26 A schematic illustrating phenomenon of 'birefringence', the characteristic property of shear wave propagation in anisotropic (fractured) medium. A polarized shear wave on entering anisotropic medium with vertical fractures aligned NW-SE, splits to two orthogonal waves travelling with different velocities. The faster wave (S_2) is polarized parallel to fracture plane and the slower (S_1) perpendicular to it, across the fracture orientation. The delay time (ΔT) between arrival times of the two waves can be measured in seismic to predict fracture geometry. The faster wave indicates the fracture orientation while the amount of delay indicates fracture density (Modified after Martin and Davis 1987)

a single system of vertical fractures (open) oriented only in one direction, the faster polarized waves correlate with the strike of the fractures indicating fracture orientation. The delay time measured between the fast and the slow waves denotes the fracture density. The arrival time of the fast wave and the delay time between the fast and slow waves *(ΔT)*, thus enables to characterize the fracture geometry in the reservoir. Fractures largely influence fluid flow (permeability) and fracture definition is essential for reservoir characterization. Naturally fractured carbonate reservoirs (NFCR) occur all over the world and reservoir modeling and simulation involves considering dual porosity and dual permeability, for which fracture characterization is essential. 3D long offset wide-

azimuth P-seismic data can also detect and measure the azimuthal variations in amplitude and velocity (AVAz, VVAz) and can help characterize fractures in reservoirs.

Limitations of Fracture Prediction

While the utility of the technique to characterize fracture geometry is potentially high, there are some inherent limitations. The technique works well in reservoirs where a simple single set of near-vertical fractures exist, oriented in a single direction. These conditions may not be met in many reservoirs. In tectonically complex areas, the reservoir is likely to have two or more sets of vertical and/or inclined fractures and oriented differently, as found in many fields. This may cause shear wave splitting many times and since azimuthal velocities are controlled by intensity and orientation of fractures linked to stress direction, it can highly complicate fracture analysis of the differently polarized waves. This can severely limit the efficacy of the technique. Furthermore, shear wave splitting may also be caused by anisotropic layers in the overburden, particularly in the near-surface shallow layers prevalent on land areas and obliterate the results of fracture study carried out in the target reservoir. Added to these problems, application of the technique needs three component shear wave data and the cost of conducting a shear seismic survey may have to be factored in for cost–benefit analysis.

References

Allen JL, Peddy CP (1993) Amplitude variation with offset: Gulf Coast case studies, Geophysical Development Series, 4, SEG, Tulsa, Oklahoma.

Anstey AN (1977) Seismic Interpretation, The physical aspects, record of short course "The New Seismic Interpreter". IHRDC

Castagna P, Swan HW, Foster DJ (1998) Framework for AVO gradient and intercept interpretation. Geophysics 63:948–956

Chopra S, Marfurt KJ (2012) Evolution of seismic interpretation during the last three decades. Lead Edge 31:654–676

Downton JE (2005) Seismic parameter estimation in AVO inversion, Ph.D. thesis, University of Calgary, Calgary, Alberta.

Ensley RA (1984) Comparison of P- and S-wave seismic data: a new method for detecting gas reservoirs. Geophysics 49:1420–1431

Kerner C, Dyer B, Worthington (1989) Wave propagation in a transversely isotropic medium: field experiment and model study. Geophys J 97:295–309

Kim KY, Wrolstad KH, Aminzadeh F (1993) Effects of transverse anisotropy on P-wave AVO for gas sands. Geophysics 58:883–888

Martin MA, Davis TL (1987) Shear-wave birefringence: A new tool for evaluating fractured reservoirs. The Leading Edge, pp 22–28.

McCormack MD, Dunbar JA, Sharp WW (1984) A case study of stratigraphic interpretation using shear and compressional seismic data. Geophysics 49:509–520

Nanda NC (2017) Qualitative analysis of seismic amplitudes for characterization of Pliocene hydrocarbon sands. Eastern Offshore, India, First Break 35:39–45

Nanda NC, Wason A (2013) Seismic rock physics of bright amplitude oil sands-a case study. CSEG RECORDER 38:26–32

Ostrander WJ (1984) Plane-wave reflection coefficients for gas sands at nonnormal angles of incidence. Geophysics 49:1637–1648

Pickett GR (1963) Acoustic character logs and their application in formation evaluation. J Pet Tech 15:650–667

Rafavich F, St CH, Kendall C, Todd TP (1984) The relationship between acoustic properties and the petrographic character of carbonate rocks. Geophysics 49:1622–1636

Robertson JD, Pritchett WC (1985) Direct hydrocarbon detection using comparative P-wave & S-wave seismic sections. Geophysics 50:383–393

Rutherford SR, Williams RH (1989) Amplitude-versus-offset variations in gas-sands. Geophysics 54:680–688

Sheriff RE (1975) Factors affecting seismic amplitudes. Geophys Prospect 23:125–138. https://doi.org/10.1111/j.1365-2478.1975.tb00685.x

Shuey RT (1985) A simplification of the Zoeppritz's equations. Geophysics 50:609–614

Stewart RR, Gaiser JE, Brown RJ, Lawton DC (1999) Converted-wave seismic exploration: a tutorial. CREWES Res Rep 11.

Tatham RH, Stoffa PL (1976) Vp/Vs—a potential hydrocarbon indicator. Geophysics 41:837–849

Wang Z (2001) Y2K tutorial, fundamentals of seismic rock physics. Geophysics 66:398–412

Wright J (1987) Short notes: the effects of transverse isotropy on reflection amplitude versus offset. Geophysics 52:564–567. https://doi.org/10.1190/1.1442325

Analysing Seismic Attributes

Abstract

Attributes are properties of reflected seismic wave signals and are analyzed to reveal geologic information they carry, concealed in the images. The attributes are essentially the kernel properties of a seismic wave, i.e., the instantaneous amplitude, phase and frequency and polarity and can be called the primary attributes. Later with advent of high resolution 3D seismic volume data, several advanced and powerful attribute techniques, such as geometric attributes have evolved. Amplitude-based attributes are the most convenient to predict reliable rock-fluid parameters, and in some cases they act as direct hydrocarbon indicators. The thin-bed related 'tuning thickness' phenomenon, complex trace analysis, sweetness and spectral decomposition techniques used widely for identifying and characterizing thick and thin oil and gas sands are commonly included in primary attributes. The geometric attributes which are important and are used extensively are derived from 3D seismic volume include the coherence, curvature and dip and azimuth attributes. These primary and the geometric attributes are discussed with their geological significance and applications in reservoir engineering and reservoir management. 'Ant tracking' an intelligent technique for mapping fractures in reservoirs is also outlined.

Display of attributes such as composite frames of multi-attributes, overlays with suitable color blending play important role in visualization and interpretation of subtle reservoir characters and are included with illustrations. Limitations of attributes and their analyses are stressed.

Attributes are intrinsic properties of seismic wave signal derived from reflection data. Seismic reflection waveforms carry valuable subsurface geologic information concealed within it and extraction of attributes and their analysis provide means to retrieve these information. Attributes that can be measured on a single trace, i.e. amplitude, frequency and polarity discussed in Chapter "Seismic Wave and Rock-Fluid Properties". Conventional 2D multitrace data further provide velocity and stratal dips that are used in early stage of exploration. But 3D multitrace high resolution, high density (HRHD) seismic volume data, however, offer many additional attributes useful for reservoir delineation and characterization which include mainly the horizontal viewing seismic slices, discussed in Chapter "Evaluation of High-Resolution 3D and 4D Seismic Data". This chapter, however, deals with more attributes, advanced and evolved, focussed specifically on delineating and characterizing thin reservoirs for exploration and engineering applications. Extracted with aid of powerful softwares and appropriately analysed, the attributes help reliable estimates of thin layer properties, usually missing in conventional amplitude attribute analysis.

The workstation-based attribute analysis is an advanced interactive interpretation linked to processing and is generally used to substantiate routine 3D interpretation, improve predictions often quantitatively and help assisting in solutions to engineering problems. It is, however, important that the geological objectives are well defined so as to select specifically the relevant attributes to be extracted and analysed for the purpose. This may at times necessitate forward modelling and rigorous advanced processing. Knowledge, experience and familiarity with processing softwares and their shortcomings, ensure proper selection of the technique and apt input parameters for reliable results. The discussion below briefly touches upon some of the most commonly analysed seismic attributes and their scope of application in petroleum exploration and production with geological implications.

Seismic attributes are essentially generated from amplitude and may be categorised as:

I. Primary attributes – amplitude, frequency, phase, polarity.
II. Geometric attributes – dip, azimuth, curvature and coherency.

Velocity, the intrinsic seismic property at times included in attribute is discussed earlier and will be further addressed in the next chapter "Seismic modelling and Inversion". AVO and impedances are also considered attributes, the former dealt in the last Chapter "Shear Wave Seismic, AVO and Vp/Vs Analysis". Impedance, the other important attribute will be discussed in the next Chapter "Seismic Modelling and Inversion".

Primary Attributes

The seismic amplitude based attribute is the most convenient and widely used to predict rock-fluid parameters which also serve as a direct hydrocarbon indicator under certain conditions. Variation in amplitudes along a reflection horizon is the simplest means of providing lithology and porosity information, though several innovative and sophisticated approaches and techniques are now in use to extract and interpret other attributes to predict rock properties quantitatively. The commonly used techniques for studying primary attributes for thin beds properties include:

- Tuning thickness
- Complex trace analysis,
- Spectral decomposition which are described.

Tuning Thickness: Thin Bed Definition and Amplitude Variation

Reflections from the top and bottom of a thick bed are well resolved and are picked easily to determine the temporal bed thickness. But for thin beds, interference of reflections causes composite reflections (Chapter "Seismic Reflection Principles—Basics") and makes it difficult to estimate thickness. Referring to Widess's wedge model, the top and bottom reflections are seen clearly up to the point of $\lambda/4$ (thickness), which provides the temporal thickness of the bed (Fig. 1). At bed thickness of $\lambda/4$, the top and bottom reflections tend to merge and beds thinner than this cause interference making it difficult to estimate bed thickness. The $\lambda/4$ is the critical thickness where maximum amplitude is seen and is known as 'tuning thicknesses'. The maximum amplitude corresponds to the dominant frequency in the bandwidth and the thickness can be broadly determined from the relation $V = n\lambda$, where 'V' is the interval velocity 'n' the dominant frequency and 'λ' the wavelength (Fig. 2b). For instance, a sand reservoir with velocity of 3000 m/s and dominant frequency of 40 Hz, common on land data, would not be resolved below a thickness of ~ 19 m.

Bed measuring less than tuning thickness is defined as 'thin bed'. Referring to Widess wedge model at Fig. 1, further thinning of the bed, does

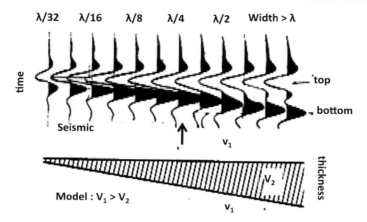

Fig. 1 Sketch illustrating 'Tuning thickness' phenomenon with help of Widess wedge model. For bed thickness more than wavelength 'λ', *top* and *bottom* reflections are resolved. As thickness reduces, the two reflections gradually tend to merge. At temporal thickness 'λ/4' (arrow marked) the amplitude shows a maxima due to constructive merger of the two reflections and is known as the 'tuning thickness'. For beds thinner than 'λ/4' interference occurs and causes composite reflections and the beds are defined as 'thin beds'. For thin beds amplitude continues to decrease till it becomes zero at end of the wedge. (Modified after Anstey 1977)

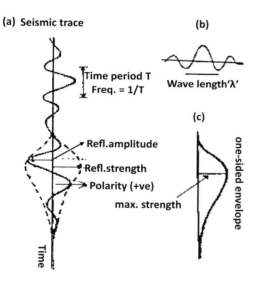

Fig. 2 Diagram showing the primary attributes of a seismic reflected wave signal (**a**) that can be measured on a trace, i.e., the reflection amplitude, reflection strength, time period, frequency, wavelength, and polarity. Reflection strength is the maximum amplitude of the envelope of a composite reflection (**c**) and independent of phase. Note in (**a**) amplitude maxima is not same as reflection strength (After Anstey 1977)

thickness for a thin bed can thus be qualitatively interpreted from its amplitude value. Widess classical model (1973) thus establishes two things: (i) occurrence of an amplitude maxima at quarter wavelength (λ/4) of bed thickness as tuning thickness and (ii) defining 'thin bed' as those lower than λ/4 that are characterized by diminishing amplitude with thinning. The thin bed dimension at a particular depth, however, will vary depending on its velocity and frequency at that depth. Nevertheless, it may be mentioned that the tuning thickness and thin bed classification based on the quarter wavelength (λ/4) criteria may vary if the model parameters at top and bottom presumed in the Widess's model change, as discussed in Chapter "Seismic Reflection Principles—Basics".

Thin beds can be detected by tuning thickness phenomenon showing high reflection amplitude, but being a composite reflection, its thickness cannot be precisely resolved directly from the conventionally processed seismic sections. This requires separate processing techniques such as 'complex trace analysis' and 'spectral decomposition' techniques described below. Incidentally, it may also be remembered that tuning thickness-related high amplitudes create impediment in AVO and other amplitude related

not show change in the waveform but shows gradual diminishing of amplitude till it becomes zero at the end of the wedge. The information on

analyses for attribute interpretation and therefore necessarily be defined.

Complex Trace Analysis—Amplitude, Frequency, Phase Polarity Sweetness

The seismic trace can be considered as a plot of particle velocity (onland data) or acoustic pressure (marine data) with time. The waveform shape of a reflection from an interface, at any time, depends on the sum of amplitude and phase attributes of reflections of individual frequencies contained in the signal bandwidth at that time (depth). The waveform for a bed also depends on thickness and impedance contrasts at top and bottom. For thin beds, however, it is difficult to assess the reflection waveforms because of nature of the composite reflection. Most seismic events in nature are composite reflections created by superposition of a number of reflections from closely-spaced interfaces. They carry vital geological information on rock-fluid properties and thickness of individual thin beds in the smeared images of summed-up response of a group of beds as a whole. As such, the reflections do not offer any clues on attributes of individual thin beds. A breakdown of the composite waveform is thus desirable to segregate the attributes of individual beds. This is achieved by transforming the time sequence to frequency domain that provides convenient means to interpret reliably the stratigraphic details. This is known as 'complex trace' analysis and is an old technique used to reveal more geologic information from amplitude, phase and frequency attributes of individual components of the composite reflection event. This can somewhat be likened to a process of obtaining derivatives that provide more and detailed information than the whole, the integrated data.

Seismic wave propagation in time is regarded as a complex signal, mathematically defined by a real and an imaginary part (Taner et al. 1979). The recorded seismic trace is the real part of an analytical signal, in x–t plane. The imaginary part, in the y–t plane, is the same sequence, but phase-shifted by 90°. The two are then summed

to construct the complex trace for extraction and analysis of individual attributes. The technique computes from a designated time window of interest, on a sample-by-sample basis and delivers the outputs; the *instantaneous amplitude*, *instantaneous frequency*, and *instantaneous phase*. The attributes are displayed commonly in colour in vertical sections to view better the geologic properties both temporally and spatially, and are generally analysed by variability in attributes rather than their discrete values.

Instantaneous Amplitude (Reflection Strength)

Reflection strength is the amplitude of envelope of an event (Chapter "Seismic Reflection Principles—Basics"). It is independent of phase, which means the maxima of the reflection strength, the one sided envelope may be different from the maximum amplitude seen at the peak or trough of the reflection (Fig. 2a, c). Reflection strength changes are considered more robust and meaningful than amplitude changes in interpretation of stratigraphic details (Anstey, 1977). Sharp lateral variation in reflection strength is indicative of major changes in rock and fluid properties that can be caused by faults, unconformities and facies changes. In some specific cases it can detect gas accumulation and delineate its lateral extent where the amplitude response due to gas abruptly changes to that of water sand. Gradual changes in reflection strength, on the other hand, may be linked to lateral changes in lithofacies and bed thickness. Though instantaneous amplitude represents the amplitude envelope of a composite reflection, its sample by sample plot showing variability may not always resolve the geologic details properly due to noise. Instantaneous amplitude is very sensitive to noise and can tend to be less reliable (Fig. 3).

Instantaneous Frequency

Instantaneous frequency is a continuous measure of the dominant frequency of a wave with time, independent of phase and amplitude. The composite reflection is comprised of reflections depending on individual frequencies contained in

Fig. 3 Showing comparison of seismic segment (**a**) and the extracted instantaneous amplitude (reflection strength) attribute section (**b**). Note the prograding sequence and the onlaps to it (shown by arrows) seen clearly in (**a**) but are poorly expressed in the attribute section (**b**) due to noise. Complex trace analysis for computing instantaneous amplitude attributes is highly sensitive to noise. (Images, courtesy by Arcis Seismic solutions, TGS, Calgary)

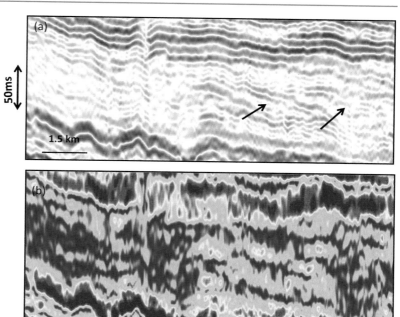

the signal bandwidth and (thin) bed thicknesses. Considering the fact that frequency is sensitive to bed thickness, the breakdown of the composite reflection, is likely to exhibit instantaneous frequency patterns that characterize the individual thin beds better (Partyka et al. 1999). At tuning thickness ($\lambda/4$), the dominant amplitude corresponds to dominant frequency and the high instantaneous amplitude is complimented by high instantaneous frequency. As the bed gets thicker the amplitude increases while the frequency decreases and conversely, when the bed gets thinner than the tuning-thickness, the amplitude decreases while the instantaneous frequency increases (refer Fig. 1). The instantaneous frequency continues to remain high as a thin bed gets thinner, with the amplitudes continuing to decrease. The instantaneous frequency complementing the instantaneous amplitude, demonstrates the phenomenon of thin bed tuning for both the frequency as well as amplitude (Robertson and Nogami 1984). This is termed '*frequency tuning*', likened to tuning thickness (amplitude).

However, deviations in frequency patterns can be due to other reasons than bed thickness. A shift toward lower frequencies known as "low-frequency shadow" is sometimes observed on reflections below hydrocarbon sands and considered a DHI indicator.. The association of frequency shadows are believed by many due to high frequency loss linked to absorption in gas. Based on energy absorption in gas, a technique known as '*sweetness*' (described below), is sometimes used by interpreters to promote high amplitude anomalies for priority drilling. Nevertheless, it continues to be a debatable issue as the causes for the shadows are not clearly understood (Taner et al. 1979; Ebrom 2004,). Absorption is a loss of energy due to frictions between the grains of a rock during propagation of a wave which is unlikely in gas molecules (Chapter "Seismic Wave and Rock-Fluid Properties"). In fact, there are many instances of gas bearing reservoirs which do not show such effects. Alternatively, high frequency loss can also be attributed to absorption in rock matrix apart from the fluid and also to transmission losses due to presence of multiple thin beds with alternating signage of reflectivity, amongst several other reasons (Chapters "Seismic Wave and Rock-Fluid Properties", "Direct Hydrocarbon Indicators (DHI)"). Nonetheless, gradual lateral variability in instantaneous frequency can be due to change in facies and such frequent and rapid variations can be typical of deposits under fluvio-deltaic environment. Analysis of instantaneous frequency attribute provides stratal details and is useful in delineation of thin hydrocarbon reservoirs, particularly the multiple thin pays embedded in thick reservoir sequence. The limit of lateral pinch-out of individual thin reservoirs can be precisely mapped from variations in pattern of instantaneous frequency changes.

Polarity

Polarity for composite reflections are difficult and only apparent polarity can be assessed. Its importance and applications in seismic well calibration and correlation and in AVO analysis for validating hydrocarbon sands are discussed in detail in Chapter "Seismic Reflection Principles—Basics".

Sweetness

Sweetness is a composite seismic attribute of amplitude and frequency and is calculated by dividing the instantaneous amplitude (trace amplitude envelop) by the square root of the instantaneous frequency (square root of average frequency). Instantaneous frequency is influenced by bed thickness and absorption and is often used in exploration to indicate relatively thicker and cleaner sands embedded within shales and are likely to be potential hydrocarbon bearing as in cases of channel sands and deltaic deposits. Essentially, sweetness for a high impedance thick sand reservoir encased in low impedance shale is indicated by strong instantaneous amplitude and with relatively lower instantaneous frequency. The sweetness is primarily driven by impedance contrasts and the higher amplitudes signify reservoir becoming cleaner or thicker while the instantaneous frequency gets lower as thickness increases. The upshot is high instantaneous amplitudes with high frequencies may indicate laminated sand reservoirs encased within shale or shale laminations within sand reservoirs.

Sweetness is a window-based attribute, and can be a quick look technique for the study of discrete sand layers in evaluating 3D volume

data for potential hydrocarbon reservoirs, somewhat similar to horizon slices. Sweetness works for those areas where considerable contrast between the rocks exists and essentially is an indicator of clean and or relatively thicker 'thin' sands and does not necessarily indicate hydrocarbon. However, when saturated with gas, the amplitude is likely to increase further, implying the high amplitude anomaly as 'bright spot' and needs validation by AVO analysis.

Instantaneous Phase

Phase and polarity, sometimes loosely considered the same are separate attributes. An instantaneous phase section displays the phase of a reflection wave form at the time corresponding to peak, trough and zero crossing. Polarity of a reflection, on the other hand, is an indicator of the signage of the impedance contrast at the interface, either positive or negative (Chapter "Seismic Reflection Principles—Basics"). The instantaneous phase is independent of amplitude, does not vary with strong or weak reflections and conveys lateral continuity of reflection events.

Often hydrocarbon reservoirs are capped by thin layers of cap shale that facies out laterally thus limiting the areal extent of trap in the structural closure of the prospect. Phase correlation can be extremely useful in tracking continuity of cap/reservoir facies where amplitude correlation does not help due to poor impedance contrasts. Instantaneous phase sections emphasize lateral discontinuities of features like faults and pinch-outs better. In poor reflectivity areas, where seismic sequence identification and facies analysis through amplitude correlation may be difficult in conventional stack, instantaneous phase sections, may be helpful in revealing better the discontinuities and the discordances in the angular patterns of reflections. Important examples of this can be picking up of subtle toplaps and shingle prograding clinoforms, commonly considered potential exploration play and often not clearly perceptible in normal sections. Use of apt choices of colour in instantaneous phase section display is likely to augment visualization of subtle features such as bed discontinuities, small faults and facies changes for better reservoir characterization than that realized from a conventional section.

However, the earlier complex trace analysis technique for extracting instantaneous attributes yields results, which are based on variability of properties, are qualitative and less dependable. Recent advances in techniques dispensing more reliable and effective results like spectral decomposition, is presently more in use for reliable evaluation of instantaneous attributes.

Spectral Decomposition (AVF)

As mentioned earlier, composite reflections are caused by interference of reflections from several closely spaced interfaces by each of the individual frequency present in the seismic signal bandwidth. Under the circumstances, it is hard to define the geometry of individual thin beds from a collective, averaged seismic response, unless the component reflections are segregated. Spectral decomposition is such a segregation technique in frequency domain, somewhat similar to complex trace analysis but much evolved and advanced. It provides output at each temporal sample of a trace to help estimate more reliably the stratigraphic details of thin beds including layer(reservoir) thickness by studying amplitude and phase variations with frequency (Castagna and Sun 2006). The amplitude spectrum provides the variability in temporal bed thickness while the phase spectrum reveals lateral extent by the discontinuity, signifying facies variations (Partyka 2000). However, it is more common with interpreters to analyse more conveniently the amplitude spectrum with frequency. The study of amplitude variation with frequency through frequency slices may be termed as *AVF* (amplitude variation with frequency), analogous to AVO.

The technique is based on the premise that each individual thin bed has a natural frequency of its own, at which it is imaged best by showing an amplitude maxima. Essentially, spectral decomposition may be considered as an extension of 'frequency tuning' concept, elevated to a refined higher level of processing for an

analytical solution to quantify thickness of thin beds. The composite reflection offers averaged amplitude as a response to frequency bandwidth, and if the amplitude is processed for a range of different frequencies, the response amplitudes will show variance. At a particular frequency, the response amplitude will show a maximum value which determines the tuning thickness for that thin bed and helps determine the layer thickness. The thinner the bed, higher is the frequency at which tuning thickness is likely to cause the amplitude maxima. From known benchmarks carried at wells, the range of amplitudes can now be calibrated to quantitatively estimate bed thickness. Spectral decomposition process creates a series of frequency slice maps, each showing amplitude response corresponding to the particular frequency (bed thickness) or frequency range (Fig. 4). The figure illustrates clearly the significant differences in the patterns imaged with bandwidth and those with individual frequencies and thus to interpret geology of thin subsurface stratal layers. For instance, referring to Fig. 4c, the interpreted channel and fan complex is imaged exceedingly well in 40 Hz frequency whereas it is not perceptible in normal seismic section imaged with 8–50 Hz bandwidth.

Spectral analysis, in frequency domain, can be accomplished by different ways of transformation and each has its advantages and limitations. It is not a unique process as a single seismic trace can have an assortment of time–frequency analyses (Castagna and Sun 2006). Nevertheless, frequency-stripping to individual frequencies results in enhancement of seismic resolution, vertical as well as lateral. It provides better vertical (temporal) resolution of bed thickness ($\sim$5–10 m, compared to usual 15–20 m) and spatial resolution of areal extent superior to conventional amplitude horizon slices. Spectral decomposition helps reveal thin bed geology with stratal details of facies such as channel, levee', point bars and crevasse splays in a geologic play of channel-fill and fan complex that could be obscured or totally missed in the normal vertical sections. The lateral variation in facies and

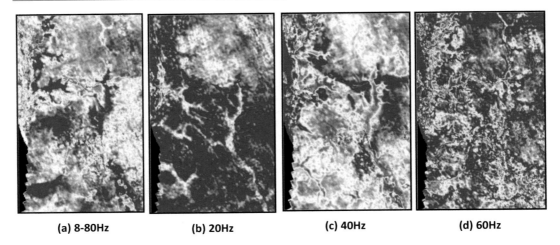

| | | | |
| (a) 8-80Hz | (b) 20Hz | (c) 40Hz | (d) 60Hz |

Fig. 4 Illustrating spectral decomposition of thin beds by displaying stratal slices imaged with bandwidth frequency (a) and by different individual components of frequencies (b, c and d) in plan view. A thin bed exhibits varying amplitudes depending on the frequency by which it is imaged and shows maximum amplitude for a particular frequency known as 'tuning frequency' analogous to 'tuning thickness'. Note the channel and fan complex feature is best imaged by 40 Hz frequency which determines its temporal bed thickness (Images, courtesy by Arcis Seismic solutions, TGS, Calgary)

thickness, mapped and overlain would help decide optimal drilling locations for maximum thickness of reservoir for hydrocarbon production. Significant applications of spectral decomposition in hydrocarbon exploration and exploitation may be as below:

- Exploration for subtle stratigraphic prospects.
- Characterization of multiple thin pays in a reservoir unit.
- Deciding drilling locations for optimal pay thicknesses for better productivity.
- Understanding reservoir heterogeneity and linked fluid-flow patterns vis-a-vis the flow units.

Geometric Attributes

These attributes are so named because they define the geomorphology of subsurface structural and stratigraphic features including their lateral variations. These are very useful attributes which help in improved and reliable interpretation of subsurface geology, especially in reservoir characterization.

Dip and Azimuth

The dip and azimuth of a bed defines its magnitude and direction with respect to a reference. 2D seismic data with no/insufficient migration, show apparent dips along the recorded profile and without azimuth information. In contrast, 3D volume data processed with prestack migration provide the true (seismic) dip and azimuth of beds. Computed from small time windows, the local lateral time gradients and their bearings derived from samples above and below, indicate the dip and azimuth of strata (Fig. 5). Analysis of dip-azimuth attributes leads quantitative estimates of size and shape of geologic bodies and their continuity and trend. Sharp discontinuities in dip-azimuth slices help indicate presence of small-scale faults and probable fracture corridors that can have significant implications during hydrocarbon production. The dip map best reflects the faults when the fault plane dips are conspicuously different than that of the horizon's dip such as antithetic faults. Azimuth maps, on the other hand, define accurately the fault plane orientations when they are opposite to the horizons (Rijks and Jauffred 1991).

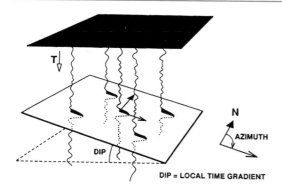

Fig. 5 Schematic illustrating extraction of dip-azimuth attributes from 3D seismic volume data. From small time windows, the lateral time gradients and their bearings derived from samples above and below, indicate the dip and azimuth of strata. Correlations are data driven and do not require manual correlation which may be laced with bias (after Rijks and Jauffred 1991)

The dip-azimuth attributes are extremely useful for 3D volume – based interpretations where horizon correlations are made automatically driven by data in the workstations. Seismic sequence stratigraphic interpretation (SSSI), discussed earlier (Chapter "Evaluation of High-Resolution 3D and 4D Seismic Data") and many other attributes need correlated seismic horizons as inputs and correlations prepared either in manual or computer auto-pick mode which may have the subjectivity based on interpreter's bias. Correlations made by dip-azimuth in 3D volume are data-driven and do not suffer from personal bias and are considered more reliable. The volumetric 3D data analysis for dip-azimuth is fast and accurate, which allows a large volume of data to be easily interpreted More significantly, dip-azimuth attributes being self-independent by themselves, do not require pre-picked seismic horizons and avoids any implanted bias (Chopra 2001).

The technique helps create, conveniently and comprehensively the tectono-stratigraphic framework of a basin, the essential input required for 3D basin petroleum system modelling. More importantly, the dip-azimuth vertical section displays can enhance resolution to pick discordant patterns implying parasequences within a seismic sequence, for targeting potential

thin plays in exploration stage. Parasequences can also be helpful in reservoir characterization in solving fluid flow problems during development and production stage. Parasequences are thin beds, bounded by marine flooding surfaces that signify paraconformities (Chapter "Seismic Interpretation Methods") which are small unconformities parallel to strata resembling bedding planes. Nonetheless, the paraconformities can influence flow units within the reservoir that can inhibit vertical communication causing anomalous flow patterns in hydrocarbon production under depletion or during water injection in secondary recovery stage.

Curvature

Curvature is a measure of how bent a surface is at a particular point; more the bent larger is its curvature. Mathematically, it is the rate of change of dip/azimuth of a surface which, in a simple way, can be represented by arcs of circles with different radii and at any point. The reciprocal of the radius offers the curvature value. The shape of a two dimensional reflector can be defined locally from the curvatures determined from two orthogonal circles of different sizes, one small and the other large, tangent to the reflector (Chopra and Marfurt 2006). The smallest circle has the maximum curvature and the biggest, the minimum. Positive curvatures conventionally signify antiforms and negative the synforms, whereas zero curvatures signify planar surface (Fig. 6). Consequently, highly deformed tight anticlines and synclines in intensely tectonized zones will exhibit higher curvature values while horizontal beds, homoclines and flanks of structures will show zero curvature. Curvature is often computed from correlated seismic horizons. It is relatively insensitive to waveform and is independent of orientation of the surface but is sensitive to horizon tracking accuracy which may be due to noise and poor data quality (Sigismund and Juan 2003). However, this concern has been alleviated to a large extent with advanced techniques of volumetric estimation of curvatures from 3D data. The technique does not require

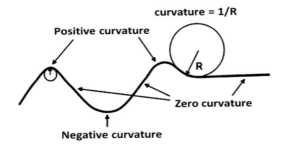

Fig. 6 Diagram illustrating 2-D curvature which is the inverse of radius of curvature and the attributes for different structures. Higher the curvature more is the surface curved. Conventionally, positive curvatures signify antiforms and the negative the synforms. Zero curvatures signify planar surface, flat or dipping such as flanks of structures. (After Chopra and Marfurt 2007)

picked horizons with its related drawbacks (Al-Dossary and Marfurt 2006).

Curvature attributes are second-order derivatives and so can enhance subtle features of folds, flexures and mounds that may be difficult to see using first-order derivatives such as the dip-azimuth attributes (Chopra and Marfurt 2007). Delicate deformations like down warps and flexures in 'karsts', faults and unconformity surfaces can be clearly observed in curvature attribute coloured-displays in horizon slices. Karsts are warped weathered irregular surfaces of carbonate rocks due to dissolution effected by water. Faults at the slippage edge would subtly form a small warp towards the down thrown side and are generally picked by positive curvatures. A chair car display planview with vertical seismic section is desirable for quality check to interpret the geologic features confidently (Fig. 7). On the other hand, gentle features like channels with flat fills may not be revealed well. An interesting application of this can be the evaluation of trough-fill prospects for hydrocarbon exploration. Curvature displays for trough-fills can be conveniently viewed in horizon slices to discriminate the interesting potential progradational/chaotic mound-fills (sand) from the continuous and flat clay-fill channels (Chapter "Seismic Interpretation Methods"). Volume-based curvature analyses being more reliable are used for indicating fracture lineaments and trends and faults with petty throws, which are not perceptible in normal stack sections due to resolution. The significance of delineating faults and fracture swarms which influence fluid flow during primary production/water injection stages in production is already overstated.

Coherence

Coherence (coherence coefficient) is a measure of similarity between two waveforms and may be determined by several means. The two simpler

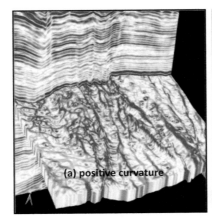
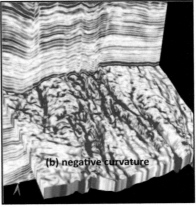

Fig. 7 Chair display of Curvature horizon slices in plan view with vertical section. The vertical sections allow review the features vis-à-vis vertical sections for confidence before interpretation. Positive (**a**) and negative curvature (**b**) attributes precisely delineate subtle high and low structural flexures. Positive curvatures are used for mapping faults (Images courtesy of Arcis Seismic Solutions, TGS, Calgary)

and common ways of coherence computation are the cross-correlation and semblance methods. Both methods essentially are based on amplitudes, though exact mathematical treatments to derive the similarity are different. However, cross-correlation technique being more sensitive to noise, the semblance is the most commonly practiced method in the industry. Seismic reflection waveform shapes depend on amplitude, frequency and phase, controlled by rock-fluid properties and thickness of layers. A change in waveform, detected by coherency would thus signify the changes in stratal properties. Coherency slices viewed in plan view offer better resolution to analyse faults and stratigraphic details which the amplitude slices may not provide.

The coherence method comprises of a pilot waveform (average), computed from a number of traces in a selected data window (Fig. 8), which then is used to measure the coherency of waveforms in data with respect to the pilot trace (Chopra 2001). Highly coherent waveforms signify lateral continuity of reflections, whereas poor coherency implies discontinuity, which are caused by unconformities, faults and fractures and facies changes. Coherence attribute is

essentially a lateral measure of similarity in a reflection horizon, highlighting the discontinuities and is also known as discontinuity or variance attribute.

Earlier, coherence attribute technique, similar to curvature attribute, required correlated horizons, liable to individual bias but with 3D data volume interpretation, the attribute is computed fast and accurate avoiding the shortcomings of pre-correlated horizon. The volume extraction delivers a coherence cube similar to dip-azimuth and curvature attributes from which time or horizon slices can be conveniently cut and displayed in planview as desired. Areas of poor reflection similarity denote low coherence (high variance) and represent discontinuities displayed usually in black to be prominent. Since ccoherence is computed in three dimensions, faults and fracture zones, stratigraphic features such as channels, deltaic deposits, turbidites, and slope and basin floor fans are revealed well. Faults, importantly the subtler ones, are easily identified in coherence slices which may not be perceptible in the conventional vertical section. A 'chair' display of coherence horizon slice and the vertical section is shown in Fig. 9. Ccoherence slices in particular, reveal the faults and fractures,

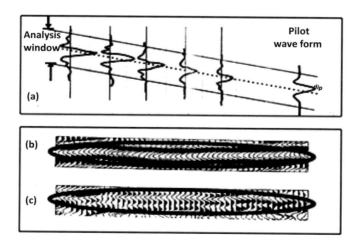

Fig. 8 Illustrating computation of coherency attribute from 3D seismic volume. (**a**) The pilot wave form is computed by averaging waveforms in a desired window from a number of traces. Semblance is computed by comparing the pilot with all the traces in the volume. Lateral continuity of similar waveforms denotes high coherence (**b**) and dissimilarity as poor coherence of vents. Horizon and horizontal slices can be taken from the computed coherence cube as per choice and displayed in plan view. (After Chopra and Marfurt (2006) and (2007))

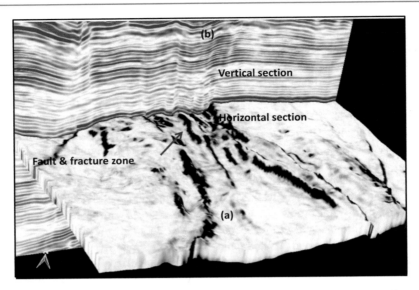

Fig. 9 Chair display of coherence horizon slice (**a**) with seismic vertical section (**b**). White portion in the plan view corresponds to good reflection events on vertical section having high coherence. The black indicates the poor/no reflection regions having low coherence and signifies discontinuities as faults and fractures The chair display of vertical section confirms the major faults. (Image courtesy of Arcis Seismic Solutions, TGS, Calgary)

oriented in in any direction including strike-slip offset faults, in contrast to conventional time slices which detect faults oriented perpendicular to the strata and may not show the strike parallel faults (Fig. 10). Stratigraphic features, though are detected by conventional horizon slices, they may lack the sharpness of boundaries that coherence horizon slice shows. This is because the discontinuities, conspicuous by low coherence, define more precisely the edge of

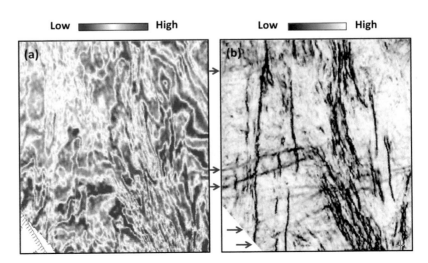

Fig. 10 Showing comparison of seismic time slice from (**a**) and coherence time slice (**b**) in fault mapping. Notice, the clear faults and fracture details seen on the coherence slice in (**b**), the black lines, which are not so obvious on the seismic slice (**a**). Note particularly the cross faults, marked by arrows, seen clearly in coherence that are incomprehensible in time slice. (Image courtesy: Arcis Seismic Solutions, TGS, Calgary)

Fig. 11 Shows comparison of mapping stratigraphic feature by seismic slice (**a**) and the corresponding coherence slice cut from coherence volume (**b**). Note the clarity in defining the edge of the river channel and the banks (**b**) over the conventional seismic slice (**a**). (Image courtesy, Satinder Chopra, Calgary)

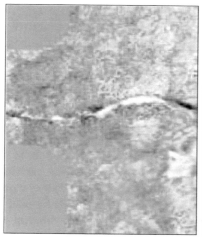 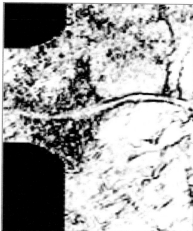

(a) Seismic horizon slice **(b) Coherence horizon slice**

boundaries, and is exemplified for a stratigraphic channel, revealed much better in coherence slice than in the amplitude slice (Fig. 11).

Ant-Tracking

Ant tracking is a very powerful imaging technique that enhances fault edges and joins the faults tracked by programmed intelligence algorithms in a curvature or coherence attribute volume. The technique, essentially based on identifying discontinuity, is very sensitive to noise and the input data requires careful reconditioning by structural smoothening filter. The structural filter emphasizes the amplitude and the structural continuity of events guided by the local structure but avoiding smoothening across faults. In a coherence volume, the technique basically takes into consideration all spatial discontinuities in three dimensions and joins the adjacent poor coherence zones by smart tracking.

The ant-tracking technique of tracing the fault surfaces and joining them is based on the idea of swarm intelligence in ant-colony systems where ants find their food source and then attract others by emitting a chemical substance called pheromone. The ant-track technique, deploys many electronic intelligent ants at different locations in the discontinuity volume to search for edge discontinuities within preset boundary conditions. The ants start search in three dimension with set parameters and captures only the features that are likely to be faults avoiding non-structural features such as channel and other stratigraphic events (Basir et al. 2013). The surfaces traversed by the swarm of ants following electronic pheromone are recorded with their orientation and strength (ants) and used to extract the fault plane surfaces. The extracted surfaces, however, are essentially segments and not complete surfaces. Joining the segments into surfaces comprises the final phase of the ant-tracking method, where fault segments are segregated into separate systems defined by their orientation and with no segments intersecting (Chopra and Marfurt 2008).

Limitations of Ant-Tracking

The highly intelligent software is quite sensitive to type of seismic data quality as it essentially tries to enhance low coherence zones which are characteristically associated with poor and no reflections, noise, unconformities and edges of stratigraphic features. The data therefore needs to be optimally reconditioned with suitable signal processing to eliminate acquisition footprints and other noises to deliver reliable coherence cube. Acquisition footprints degrade seismic attributes and are particularly troublesome in mapping faults and fractures. The efficacy of the ant track technique is thus highly influenced by the quality of the input coherence or curvature cube. Depending on the geology of the area and the fault attributes the data may have to undergo suitable filters to provide superior and noise free seismic input. The technique also depends on setting the ant-track search parameters judiciously which warrant comprehensive understanding of the area geology, the seismic data processing and the application of the software. Merely relying on the technique for characterizing the reservoir fracture geometry to solve the problems, especially linked to reservoir development and production can be significantly misleading.

Multiattributes Analysis and Composite Coloured (Overlay) Displays

Multiattributes Analysis

Though attributes are useful in both exploration and production related issues, it is in the later stage of reservoir management, the multiattributes analysis can provide significant value addition to field development plans and reservoir management. Multiattributes analysis is essentially an interactive processing and interpretation method. For reliable solution to critical reservoir and production related problem, it needs a synergetic and systematic approach to work flow, broadly a three-step process. First a feasibility study may be made by understanding the rock-fluid properties from the petrophysical studies of the well logs *vis-a-vis* the seismic data quality at hand. This is followed by seismic rock physics modelling factoring all relevant information from the multidisciplinary data sets such as geologic, petrophysical, petrography and reservoir and production engineering prior to attempting multi-attribute analysis. This helps in choosing the attributes most sensitive to the problem to be used along with suitable available softwares and amenability of seismic data quality as input. In most cases, the data may have to be

reconditioned which needs careful application of interactive processing with optimum parameters and advanced techniques. And finally, the most vital part, the analysis and synthesis of the multiattributes outputs for arriving at solution that converges to satisfy all or most of all data from different disciplines.

Interactive processing and interpretation of seismic attributes is a real challenge to the skill and knowledge of the seismic analyst. The expertise and experience of the data processing expert matters as much as that of the interpreter, in conditioning data which includes noise cancellations, removal of acquisition footprints, correcting for zero phases, broadening the spectral bandwidth and preserving the true amplitude. A rigorous recalibration of seismic with wells is most critical in the workflow prior to attribute computations. Selecting display modes, scale and appropriate colour blending can sometimes be strikingly revealing and needs to be considered properly.

Composite and Overlay Colour Displays

In the context of what has all been stated, choice of composite display mode and proper colour blending of attributes can be important in perceiving more information to evaluate reservoir geology with better confidence. Colour blending promotes visualization of geologic events due to increased optical resolution. Composite coloured displays of more than one attribute greatly help in synthesizing evaluation of the geologic and engineering problems particularly during field development and production stages for better reservoir management.

Multiattributes, Applications and Significances

Composite display of multiattributes, suitably colour-blended can be a powerful tool useful in reservoir characterization (Fig. 12). The coherence stratal slice (Fig. 12a) delineates a river

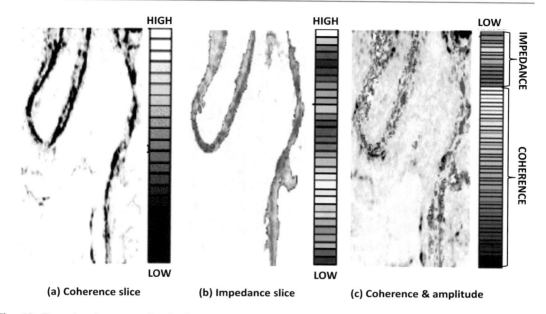

(a) Coherence slice   **(b) Impedance slice**   **(c) Coherence & amplitude**

Fig. 12 Example of a composite display of attribute slices with color blends depicting channel, (**a**) coherence (**b**) impedance and (**c**) the composite overlay of impedance and coherence slices The color bars are indicated. Notice the important role of color blending in impedance attribute (**c**) which shows appreciable variation in low impedance values not perceived in (**b**) which can have geologic significance for the reservoir facies (image courtesy: Satinder Chopra, Calgary)

channel while the impedance slice (Fig. 12b) exhibits the associated facies mostly of low and medium impedance values (sands and shales). Over-lay of the two attributes and given graded colour-blending (Fig. 12c) shows improved optical resolution resulting in perceiving variations in impedance values that are not so well seen in Fig. 12b. This is noteworthy, as smart colouring can help discriminate the reservoir facies within the channel that can have significance in reservoir characterization. Another example of a coherence slice imaging a typical distributary channel and a composite display of coherence and amplitude strength reveals the high amplitude patches which may be inferred as channel and the associated 'crevasse play' sands beyond the channel banks (Fig. 13). However, to infer the sands as gas bearing would need conformation by other attributes such as *Vp/Vs* and AVO. Yet another interesting illustration of composite coloured multiattributes slices that can potentially impact fluid flow in reservoir is shown in Fig. 14. For instance, consider the

amplitude slice in Fig. 14a in the figure delineating the representative hydrocarbon sands, while the coherence slice Fig. 14b defines the geometry of numerous faults (black lines) affecting the reservoir. The amplitude overlain on coherence slice Fig. 14c displays the likely impact of the faults on fluid flow pattern in the reservoir not anticipated and considered during static reservoir modelling and simulation. The faults may act as barriers leading to anomalous fluid flow during depletion or block watersweeps during water flooding in secondary recovery processes. Multiple faults and their network in the reservoir, detected from amplitude and coherence slices by composite overlay display can have also other significant consequences. The faults can also affect reservoir connectivity and create compartmentalization with fluid flow controlled by fault-permeability. These are extremely vital production issues that influence efficient reservoir management.

In general, seismic attributes work well in good signal-to noise data which degrades with

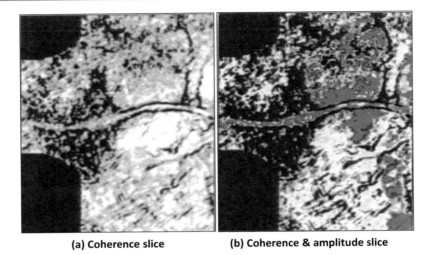

(a) Coherence slice   **(b) Coherence & amplitude slice**

Fig. 13 Another example of composite display-coherence attribute slice depicting a channel (**a**) and coherence slice overlain by colored amplitude slice (**b**). Note the variations in color shades which may denote reservoir facies variation associated with channel sands.

The high amplitude portions (red) are likely to be hydrocarbon bearing channel and crevasse splay sands which otherwise would have been difficult to predict by analyzing the attributes singly. (Image courtesy, Satinder Chopra, Calgary)

(a)   **(b)**   **(c)**

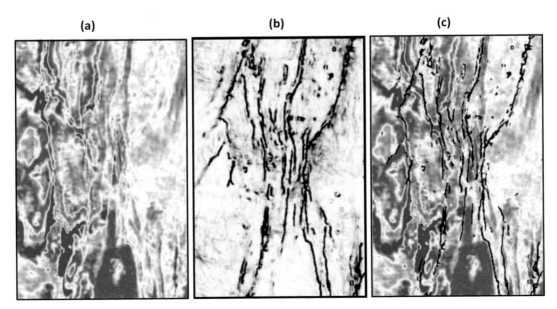

Fig. 14 Plan view display of multiattributes illustrating significance of colored composite and overlay of attributes slices. Such displays are helpful in better visualization of reservoir geology and offering solutions to reservoir management problems. For instance, consider the amplitude time slice (**a**) showing hydrocarbon sands, and the coherence time slice (**b**) the network of faults

(black lines). The amplitude overlain on coherence slice (**c**) displays the likely impact of the faults on fluid flow in the reservoir. The faults may act as barriers leading to anomalous fluid flow during depletion and water flooding in secondary recovery processes (Images courtesy, Satinder Chopra,Calgary)

depth. Attribute slices at shallow levels are a good quality check for the data quality and may be precursor to judge the expected quality of attribute to be extracted from data at depths for interpretation. The geometric attributes work exceptionally well in marine data which generally have good signal–noise ratio and high resolution. In seismically poor areas, particularly in land data, and at greater depth, the curvature and specifically the coherence attribute efficiency may be impacted, susceptible to accuracy in reflection correlation even though performed by computer. Dip-azimuth, curvature and coherence attributes, analysed jointly, facilitates understanding subtle deformations and discontinuities within a reservoir that can greatly help mitigate reservoir related engineering issues during field development and production. Apparently innocuous minor offset faults, facies changes and permeability barriers such as shale strings can cause severe reservoir heterogeneity which if unnoticed and not factored in, may significantly influence fluid flow patterns impeding production performance. Display of multiattributes slices in composite and overlay modes with suitable colour blending in plan-view and evaluated along with seismic vertical sections in the backdrop is desirable for dependable evaluation of attributes.

Limitations in Attribute Studies

To have meaningful and reliable geological and engineering information from analysis of seismic attributes, presence of reasonably noise-free reflections with good resolution is desirable. The seismic data is often of lower bandwidth which would require augmentation of frequency contents to the utmost of the range recorded. The expertise and experience of the data processor matters in proper data conditioning which includes noise cancellations, removal of acquisition footprints, correcting for zero phases, broadening the spectral bandwidth and preserving the true amplitude. Judicious choice of display mode, scale and colour blending can sometimes be revealingly important. Ultimately, lot depends on the agreeability of data quality to

its object-specific processing modalities and it must be ascertained before starting the analysis.

Seismic response also depends largely on the depth of occurrence and the geologic environments which influences the rock-fluid properties. Responses are expected to vary widely, under different depositional and tectonic settings. Even within a limited small depth range in a well, sand and shale units can have dissimilar rock properties and different seismic responses. Attribute studies require appropriate calibration with well data for reliability to be effective. Attributes are easy to extract but can be challenging to interpret and evaluate. Extraction of multiple attributes may be trivial but can be difficult to interpret when they imply contrarian results. The seismic attributes under diverse geologic scenarios may deliver a variety of outputs, of which some may be mutually complimentary and others opposing which inhibits an unambiguous convergent solution. Attribute interpretation is not unique and interestingly, image displays put in any form can be visualized to have some viable geologic pattern depending on the interpreter's perception and visualization. It is important the inferences drawn are checked for veracity to be in agreement with the well log, geologic and engineering data. For instance, an attribute pattern visualized as a feasible channel when verified by overlaying it on paleostructural map, could reveal the channel geometry and flow direction against the regional dip, which is flawed.

The interpreter needs to understand the specifics of the attribute processing methods, the assumptions in their built-in logics and more importantly, their suitability to the geologic setting of the area before selecting the particular attribute analysis in the work-flow. If the attribute pattern cannot be evaluated as solutions linked to specific exploration or development problems at hand, the exercise can be pointless. Attribute analysis without geological support and done in isolation can be regressive, leading to pitfalls. It is a conscientious collaborative task and not merely one of applying software to provide some attribute patterns and values, especially in issues concerning reservoir management where the seismic technique's reputation and management's stakes can be high. Working familiarity with the various techniques and experience of an interpreter and more importantly one's co-ordination with the processing expert can go a long way to insure effective results and indeed can be the true tribute by the geoscientist to seismic attribute.

References

Anstey AN (1977) Seismic interpretation: the physical aspects. IHRDC, p 3–1 to 3–19 and 3–54–85

Basir HM, Javaherian A, Yaraki MT (2013) Multi-attribute ant-tracking and neural network for fault detection: a case study of an Iranian oilfield. J Geophys Eng 10:p3-9

Castagna JP, Sun S (2006) Comparison of spectral decomposition methods. First Break 24:75–79

Chopra S (2001) Adding the coherence dimension to 3D seismic data. CSEG Recorder 26:5–8

Chopra S, Marfurt K (2007) Seismic curvature attributes for mapping faults/fractures, and other stratigraphic features. CSEG Recorder 32, view issue

Chopra S, Marfurt KJ (2008) Emerging and future trends in seismic attributes. Lead Edge 27:298–318

Chopra S, Marfurt KJ (2006) Seismic attribute mapping of structure and stratigraphy: distinguished instructor short course. SEG and EAGE, Tulsa, pp 1–131

Dossary-Al, Marfurt J (2006) Geophysics 71(5):41–51

Ebrom D (2004) The low-frequency gas shadow on seismic sections. The Leading Edge 772

Partyka G (2000) Seismic attribute sensitivity to energy, bandwidth, phase and thickness. In: 70th annual international meeting, SEG, Expanded Abstracts, pp 183–186

Partyka G, Gridley JM, Lopez J (1999) Interpretational applications of spectral decomposition in reservoir characterisation. Lead Edge 18:353–360

Rijks EJH, Jauffred JCEM (1991) Attribute extraction: an important application in any detailed 3-D interpretation study. Lead Edge 10:11–19

Robertson DJ, Nogami HH (1984) Complex seismic trace analysis of thin beds. Geophysics 49:344–352

Sigismondi ME, Soldo JC (2003) Curvature attributes and seismic interpretation: case studies from Argentina basins. Lead Edge

Taner MT, Koehler F, Sheriff RE (1979) Complex seismic trace analysis. Geophysics 44:1041–1063

Widess MB (1973) How thin is a thin bed? Geophysics 38:1176–1180

Seismic Modelling and Inversion

Abstract

Modelling is a computational process to determine the seismic response of a given geologic model which can be structural or stratigraphic and done in one, two or three dimensions. Synthetic seismogram is the most elementary and common example of one-dimensional (1D) modelled seismic response of an earth model whereas a synthetic non-zero offset VSP is two-dimensional. Application, benefits and limitations of forward modelling processes are briefly stated with examples.

Inverse modelling or 'inversion' is the reverse process of forward modelling where the subsurface geologic model is synthesized from the seismic data. Operator-based seismic inversion transforms seismic reflection amplitudes to inverted reflectivity or impedance of the subsurface earth series. This provides an effective and powerful tool to interpret seismic data in terms of geology based on impedance. Impedance is a layer property, in contrast to reflection amplitude which is an interfacial property and offers more geologic information about the layers. Impedance attributes obtained from inversion process have improved resolution and provide more reliable rock and fluid properties which are the crucial input to reservoir modelling and flow simulation done by the reservoir engineers. This has made inversion an integral part of interpretation work flow for reservoir management.

Acoustic (AI), elastic (EI) and simultaneous inversions (SI) techniques and velocity, density, Poisson's ratio and AVO inversion are briefly discussed. Thin bed reflectivity series and its transformation to acoustic impedance that provides more resolution to characterize thin beds is emphasized with example. Role of seismic inversion in reservoir management with Inadequacies and shortcomings of the techniques are mentioned.

The surface seismic reflection method provides images of the earth's subsurface which are responses in two (2D) or three dimension (3D). Geoscientists interpret the seismic data consisting of multiple traces as they look for indication of oil and gas, a task which is often fraught with challenges. In their quest for understanding seismic data in terms of reflection arrival time, amplitude, impedance, and velocity for predicting subsurface geology, they look for tools that can help them in such exercises. Seismic modelling is a very useful simulation tool employed for such purpose. It is a computational process to determine seismic response of a given geologic model that helps in understanding the intricacies of actual field seismic images. The propagation of seismic waves from the source to the subsurface lithological boundaries and their reflections back to ground receivers are simulated to obtain a synthetic seismic section. The process of computing the response is known as seismic

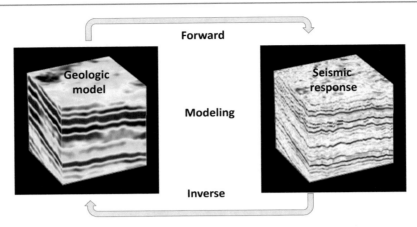

Fig. 1 A conceptual diagram illustrating seismic model-ing. Forward modeling generates the seismic response from a given geologic model whereas inverse modeling generates a geologic model from a given seismic response. *(Image courtesy of Arcis Seismic Solutions, TGS, Calgary)*

forward modelling. Conversely, the reverse pro-cess of deciphering the geologic model from the observed seismic data is termed inverse mod-elling (Fig. 1).

Physical and Numerical Modelling

Seismic responses can be simulated for models that are physical, as in laboratories, or computed mathematically (numerical), where the rock properties are specified as model inputs. Though laboratory physical models are real and can be made to imitate the earth model, the outputs are static response of rock properties and may not replicate the dynamic response as the actual seismic image would record on the ground. There are also issues involved primarily because of simulating the subsurface geology in a labo-ratory would be much different from the *in situ* natural set up in the subsurface. The most chal-lenging factor, perhaps, is the huge order of inequality in dimensions of the models involved, in the laboratory and in the nature (Chapter "Seismic Wave and Rock-Fluid Properties"). Numerical modelling allows the computed responses to be reviewed and with options for

tweaking parameters for recalculating response for a reasonable match. The modelling process is fast and convenient and consequently, numerical modelling is more common in practice than physical modelling.

Simulating the response of subsurface geology through modelling serves several purposes for application in diverse fields of seismic acquisition, processing and interpretation (API). Forward modelling is widely used in seismic data acqui-sition for optimising survey lay out designs and acquisition parameters for different geologic problems. In seismic data processing modelling is used for testing and calibrating algorithms. Par-ticularly for 3D surveys, which are relatively expensive, the parameterization of source-receiver geometry for deciding mainly the optimal bin sizes, CDP foldage and multiazimuthal coverage, requires modelling to choose the best cost-effective option to achieve the survey objective.

Interpreters, typically employ seismic mod-elling to gain better insight to understanding of observed seismic response to geologic changes so as to attribute these variations assertively to rock and fluid properties. Modelling application in interpretation includes creating geological models to investigate structural and stratigraphic problems, especially in geologically complex

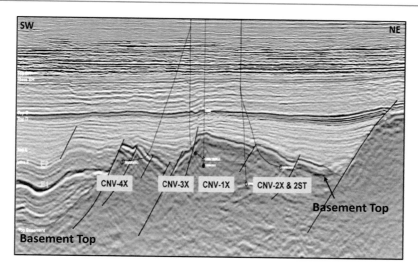

Fig. 7 Seismic section through CNV field in Cuu Lon basin, offshore Vietnam. Note the excellent quality of reflection amplitude and continuity from top of the intensely fractured basement. The criss-cross events of reflections from the fault planes can be sen which signify the fracture network. (after Koning 2014)

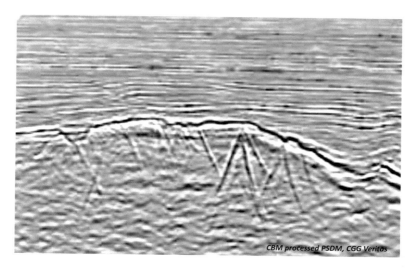

Fig. 8 Showing improved version of seismic segment through the well CNV 3X, offshore Vietnam processed with controlled beam migration (CBM). The prestack depth domain (PSDM) clearly shows the improved images of faults and fractures. Note the clarity in fault plane imaging, showing orientation and extension of faults and fractures within the basement which allows extract of seismic discontinuity attributes reliably. (After Koning 2014)

geometry, makes the exploration for this unconventional reservoir highly challenging and risky. Many basement oil/gas fields after initial success have proved to be uneconomical as in Dai Hung (Big Bear) in Viet Nam and Beruk Northeast Oil pool in Central Sumatra (Koning 2019). The setback that happened in the Dai hung field in Cuu Long basin, offshore Vietnam, is strange as it has the same tectonic and structural setting and similar reservoir characteristics as that of the giant prolific producer Bach Ho oil field. The field initially produced good amount of oil but within a span of five years production declined and field was declared unprofitable. Dai Huang (Big Bear) illustrates the complexities, uncertainties and challenges associated with basement oil and gas fields (Koning 2019). Initial acclaims of finding hydrocarbon in the basement thus can turn to frustrating experience as later wells may not find oil for sustained economic production and resulting in exploration set-backs, costing time and money.

Nonetheless, instances of world-wide occurrence of many giant oil and gas fields and easy conventional prospects becoming rarer, explorationists are swayed to pursue complex and challenging basement plays for finding new oil. The basement exploration strategy may include revisiting the depleted conventional reservoirs where initially hydrocarbon finds in basement were noticed. In giant-sized fields La Paz, Venezuela and Octongo field, Argentina, oil in fractured basements were discovered 30 and 100 years respectively after the overlying conventional sedimentary reservoirs were depleted after long period production (Koning 2019). This underscores the point that many of the basement pools that ceased production after a short period may need to be revisited for reanalysis of data. Cognitive analysis by synthesis of regional geological, seismic, well logs, cores and other borehole information would lead to understand the type and intensity of tectonic stress, the reservoir rock types and related fracture systems. Evolution of the basin, built with tectono-stratigraphic framework to assess the petroleum system modeling (BPSM) may also be important for reevaluating the basement plays. To reduce exploration risks, basement exploration may also be considered as secondary object with the overlying conventional prospects as the primary target.

Interestingly, the fractured-basement fields are mostly discovered serendipitously. Wells are drilled primarily to explore structural and stratigraphic prospects in the overlying sedimentary section as the primary target. Often when the targets are found none too encouraging and the basement is not too far deep, drilling into basement may be an attempt for a chance hydrocarbon find and as well to have additional geologic information. Exploring basement hydrocarbon generally involves enormous amount of uncertainty and discoveries cannot ignore serendipity.

References

Belaidi A, Bonter DA, Slightam C, Trice RC (2016) The Lancaster Field: progress in opening the UK's fractured basement play, Geological Society, London, Petroleum Geology Conference Series, vol 8, pp 385–398. https://doi.org/10.1144/PGC8.20

Hung DN, Le HV (2004) Petroleum Geology of Cuu Long Basin—Offshore Vietnam, Search and Discovery Article #10062

Koning T (2014) Global accumulations of oil and gas in fractured and weathered basement: best practices for exploration and production. Search and Discovery Article #20281

Koning T (2019) Exploring in Asia for oil and gas in naturally fractured and weathered basement reservoirs, Geoconvention, Calgary

Saran A, Mehta CH, Benjamin S, Verma BM (1993) Seismic detection of fractured zones in Bombay high basement. In: Proceedings of second seminar on Petroliferous Basins of India, vol 2, pp 399–405

Sengupta S, Nanda N (2011) Fractured basement reservoir: limitations of seismic evaluation. In: Conference paper, EAGE, 73rd EAGE conference & exhibition incorporating SPE EUROPEC 2011, Vienna. https://doi.org/10.3997/2214-4609.20149002

Whaley J (2016) Hiding in the basement. GeoExpro 13(6)

Younes AI, Engelder T, Bosworth W (1998) Fracture distribution in faulted basement blocks: Gulf of Suez, Egypt, Geological Society, London, Special Publications, vol 127, pp 167–190